Recent Advancements in Aluminum Alloys

Edited by Shashanka Rajendrachari

Published in London, United Kingdom

Recent Advancements in Aluminum Alloys
http://dx.doi.org/10.5772/intechopen.103993
Edited by Shashanka Rajendrachari

Contributors
Emad Toma Bane Karash, Mohammad Takey Elias Kassim, Emrah Fahri Ozdogru, Bazani Shaik, Muralidhara Rao, G. Harinath Gowd, B. Durga Prasad, J. Ranga, Omid Emadinia, Fahad Zafar, Manuel F. Vieira, Ana Reis, Helder Nunes, Ying Pio Lim, Heng Kam Lim, Olayinka Olaogun, Esther Titilayo Akinlabi, Cynthia Samuel Abima, Isaac Enesi Dongo, Monday Itopa Momoh, Fawzy H. Samuel, Ehab Samuel, Agnes M. Samuel, Victor Songmene, Hicham Tahiri, Shashanka Rajendrachari, Emre Altaş, Vutukuru Mahesh

First published in London, United Kingdom, 2024 by IntechOpen
IntechOpen is the global imprint of INTECHOPEN LIMITED, registered in England and Wales, registration number: 11086078, 5 Princes Gate Court, London, SW7 2QJ, United Kingdom

British Library Cataloguing-in-Publication Data
A catalogue record for this book is available from the British Library

Additional hard and PDF copies can be obtained from orders@intechopen.com

Recent Advancements in Aluminum Alloys
Edited by Shashanka Rajendrachari
p. cm.
Print ISBN 978-1-83768-508-0
Online ISBN 978-1-83768-509-7
eBook (PDF) ISBN 978-1-83768-510-3

For EU product safety concerns:
IN TECH d.o.o., Prolaz Marije Krucifikse Kozulić 3, 51000 Rijeka, Croatia,
info@intechopen.com or visit our website at intechopen.com.

Meet the editor

Dr. Shashanka Rajendrachari is an assistant professor in the Department of Metallurgical and Materials Engineering, Bartin University, Turkey. He obtained a Ph.D. from the Department of Metallurgical and Materials Engineering, National Institute of Technology Rourkela, India. He obtained an MSc in Industrial Chemistry and an MTech in Nanoscience and Technology from Kuvempu University, Karnataka, India. He is the recipient of numerous national and international awards for his research contributions. His research areas are nano-structured alloys, powder metallurgy, electrochemical sensors, corrosion, and smart and high-entropy alloys. He has 8 years of research and teaching experience and has published more than 60 research articles, 13 books, and 9 book chapters.

Contents

Preface

This book, *Recent Advancements in Aluminum Alloys*, consists of eleven chapters that address different topics related to aluminum alloys, ranging from the basics of aluminum alloys to advanced applications of aluminum alloys prepared by various techniques. The book is organized into four sections: "Introduction to Aluminum Alloys", "Fabrication of Aluminum Alloys", "Properties of Aluminum Alloys", and "Advanced Applications of Aluminum Alloys".

Section 1 includes two chapters. One introduces the topic and the other focuses on aluminum alloys and their grain refinement. Both chapters are important for understanding the basic principles of aluminum alloys and their properties and applications. Section 2 includes three chapters on the fabrication of aluminum alloys by various advanced techniques. The chapters discuss extrusion, low- and high-pressure casting, and additive manufacturing techniques to prepare different grades of aluminum alloys. Section 3 consists of two chapters that examine the physical and chemical properties of aluminum alloys and how the properties of aluminum alloys can be altered by adding dispersing agents. They also discuss how the composition of aluminum alloys affects their various physical and chemical properties. Section 4 includes four chapters that review advanced applications of aluminum alloys. They discuss how environmental wastes can be converted to environmental wealth by using an eggshell as reinforcement in aluminum alloys. Chapters in this section also discuss corrosion resistance applications, friction stir welding processes, and more.

Recent Advancements in Aluminum Alloys is designed to help mechanical, biomedical, and chemical engineers, metallurgists, materials scientists, chemists, and those working in industry. It is also a useful resource for undergraduate and graduate students.

Dr. Shashanka Rajendrachari
Assistant Professor,
Department of Metallurgical and Materials Engineering,
Bartin University,
Bartin, Turkey

Section 1

Introduction to Aluminum Alloys

Chapter 1

Introductory Chapter: Introduction to Aluminum Alloys

Emre Altaş, Shashanka Rajendrachari and Vutukuru Mahesh

1. Introduction

With the advancement of technology and scientific studies, new areas of use have been created for existing materials. In this way, the interest in new materials with high quality that can meet current needs has increased [1]. The developing technology and scientific studies to date are in search of improving the desired properties of existing materials and the emergence of new materials that can be alternatives. This search continues with the emergence of materials with new qualities [2]. Compatibility of physical and characteristic properties is very important in material selection. For this reason, many methods have been developed and continue to be developed today. Particularly in the aviation, space, gas turbines, automotive, and maritime industries, materials that are high-performance, light, and durable, and can combine features such as stability at high temperatures are needed [3–5].

Aluminum and its alloys, which are widely used especially in the aviation industry, can be preferred and used in long-lasting applications due to their features such as low density, high specific strength, and resistance to oxidation [6, 7]. Among the reasons why aluminum metal and its alloys have different usage areas in different sectors are that they can be easily produced, shaped, and processed, have high corrosion resistance, are lightweight, and have good strength properties. Additionally, aluminum is extremely suitable for recycling [8, 9]. Aluminum alloy is almost infinitely recyclable, and the recycling process requires only 5% of the energy of primary aluminum production. High-purity aluminum is a soft material with an ultimate strength of approximately 10 MPa, which limits its usability in industrial applications. To compete with other building materials, the strength of Al-based materials needs to be significantly increased. There are several ways to increase the strength of metallic materials: alloying with sufficient elements, adding appropriate strengthening particles, plastic deformation, or grain size reduction [10].

Aluminum alloys are the most used metallic engineering materials after steel today. Aluminum and its alloys, which are widely used especially in the aviation industry, can be preferred and used in long-lasting applications due to their features such as low density, high specific strength, and resistance to oxidation. Among the reasons why aluminum metal and its alloys have different usage areas in different sectors are that they can be easily produced, shaped, and processed, have high corrosion resistance, are lightweight, and have good strength properties. In addition, aluminum alloys provide significant advantages over other engineering materials due to their features such as high thermal conductivity, nonflammability, and being completely recyclable and weldable [11, 12].

During the production phase, material scientists constantly improve the properties of the materials they have obtained by moving them to macro dimensions under more minimized conditions. The purpose of this is that today's technology is competitive in the market and the features of the materials needed are no longer affordable. Above all else, one of the most distinctive features required from materials is to ensure continuity [13–15]. In other words, the material produced gives the same physical results and reactions when used at different times but under the same conditions. Therefore, it is important that the material has a homogeneous microstructure and an equal stress distribution on the material [16–18]. In recent years, different production methods have been developed in addition to traditional powder metallurgy methods in the production of some critical parts. Mechanical alloying is one of these methods. This method is used to alloy metals without exposure to any chemical or heat treatment [19–21]. With mechanical alloying, metals can be alloyed without the need for melting or heat treatment. The mechanical alloying process involves cold welding of solid powder particles to each other and breaking them after deformation hardening. At the same time, particle-reinforced composites are produced using the mechanical alloying method, in which the best microscopic or macroscopic combinations of different materials are provided in order to obtain fine microstructures in powders [22–25].

2. Aluminum and aluminum alloys

Aluminum is found in the form of aluminum silicate in rocks, feldspars, feldspathotites, and micas in clay soils formed by the disintegration of bronze, bauxite, and iron-rich laterite. Bauxite, the most important aluminum ore, contains 52% aluminum oxide [26, 27].

Aluminum is the most abundant metallic element in the earth's crust after iron, and it became economically produced in engineering applications toward the end of the nineteenth century. With the emergence of the first internal combustion engine vehicles, the engineering value of aluminum as an automotive material began to increase. Aluminum, which meets the need for a light and conductive material for the transmission of electricity over long distances, has also taken its place in the new industry that was born on strong, light, and break-resistant parts, along with the work of the Wright brothers on aviation. Nowadays, aluminum is used to obtain value-added products in many transportation sectors such as automotive, defense industry, aviation, rail systems, and maritime. Aluminum production is generally made from ore and recycled scrap. Today, aluminum production from ore is approximately twice that of aluminum production from scrap [28, 29]. Aluminum is the third most abundant element after oxygen and silicon, with a content of 8% in the earth's crust. Even though there are so many, it was discovered after minerals such as iron, copper, tin, lead, gold, and silver. Aluminum is found in nature as a compound. The existence of this abundant element was detected only in 1808 by the British Sir Humphry Davy [30].

The most distinctive feature of aluminum, which is used in all areas of human life and especially in engineering applications, is its lightness. It is the lightest metal after magnesium and beryllium. Due to the superior properties provided by aluminum and its alloys, their consumption is increasing rapidly and new areas of use are opening up every day. Although pure aluminum is a very active metal in the galvanic series, the protective oxide layer that easily forms on its surface makes it widely used.

This impermeable, hard, and protective oxide layer consisting of aluminum oxide (Al_2O_3) significantly increases the corrosion resistance of aluminum. Accordingly, as aluminum is purified, its corrosion resistance and conductivity increase. For this reason, aluminum alloys, which are very sensitive to corrosion, are now protected from corrosion by cladding in pure aluminum [31]. On the other hand, the very low strength of pure aluminum can be increased by cold working. Today, aluminum and its alloys have, due to its properties, become one of the most important construction and engineering materials used in the industry. While it has properties, such as high thermal and electrical conductivity and corrosion resistance in its pure form, these properties have spread to a much wider spectrum with alloying and have become widely used. Today, more than 100 aluminum alloys are widely used in industry. The most important features are summarized below [26]:

- Aluminum is lightweight. It weighs only one-third of the weight of a steel material of the same volume.
- Aluminum is resistant to weather conditions, foodstuffs, and many liquids and gases used in daily life.
- Aluminum has high reflecting ability. With its silvery white color contributing to this feature, it has an attractive appearance for both interior and exterior architecture. This beautiful appearance of aluminum is achieved by anodic oxidation (anodizing), lacquer materials, etc. with applications, such as period can be preserved. In fact, in many applications even a natural oxide layer is sufficient.
- The strength of various aluminum alloys is equal to or higher than that of ordinary structural steel.
- Aluminum is an elastic material, especially for chipless manufacturing processes. Therefore, it is resistant to sudden impacts. Additionally, its durability does not decrease at low temperatures. (The resistance of steels against sudden impacts decreases at low temperatures.)
- Aluminum is a metal that is easy to machine. So much so that it can be turned into foil or wire with a thickness of less than 1/100 mm.
- Aluminum conducts heat and electricity, as well as copper.
- All methods, such as casting, forging, rolling, pressing, extrusion, and drawing, can be applied to shape aluminum.

2.1 Mechanical properties of aluminum

The modulus of elasticity of aluminum is approximately one-third of that of steel. In other words, the amount of elastic deformation of aluminum under the same load is three times that of steel. This feature is of great importance in design calculations. The low modulus of elasticity is considered an advantage when exposed to impact loads since the resistance of aluminum is higher than that of steel. It allows the absorption of large amounts of energy [32]. One of the main features of aluminum is its ease of shaping and processing according to conventional manufacturing methods. Since pure metal is soft and has the ability to become wire, it can be rolled, drawn, and shaped by applying various cold processes such as drawing, bending, pressing, and molding [33].

2.2 Chemical properties of aluminum

Aluminum is not available in pure form due to its high chemical activity. Therefore, its production is made from bauxite ore consisting of iron oxide and aluminum silicate. Due to a fixed oxide layer formed on the aluminum surface in contact with air, the metal and its alloys generally show great resistance to the corrosive effects of the atmosphere. Aluminum reduces the oxides of other metals due to its affinity for oxygen. Due to this feature, powdered aluminum is used in the production of metal oxides such as chromium, vanadium, barium, and lithium by reducing them [10]. Since aluminum is nontoxic, it is widely used in many areas, especially in the production of equipment in the food industry. Again, due to this feature, it is widely used in the packaging of food and medicines, cigarettes, and tea [34].

2.3 Physical properties of aluminum

Low density, one of the physical properties of aluminum, comes to the fore in many applications. The density of aluminum in the commercial group is approximately 2.70 g/cm^3. When equal volumes are compared, aluminum has approximately one-third the weight of iron, copper, and zinc. In some applications, it is not enough to focus on the lightness advantage of metal alone. For example, a material that does not have sufficient strength but has low density is not very useful for the elements that make up the structure of an aircraft. In this case, while pure aluminum is not suitable for use, aluminum alloys are used, where strength and lightness are desired together [5].

Aluminum and its alloys also need different criteria in comparison with the traditional materials and manufacturing methods with which they have to compete. When expressed in terms of concepts such as specific strength, specific stiffness, and discontinuous yielding during forming, aluminum alloys exhibit equivalent and sometimes superior performance compared to traditional materials [2].

These properties of aluminum make it preferred for the automotive and manufacturing sectors. Saving fuel and reducing costs due to its lightness in the transportation sector and regulations on the emission number of vehicles on national and international platforms have made aluminum the best alternative material for the transportation sector [35].

2.4 Usage areas of aluminum alloys

Increasingly complex production methods, differentiating and diversifying consumer demands, increasing population, and the awareness that the limits of natural energy resources are rapidly approaching over time with production have necessitated radical changes in the production methods and raw materials that have been customary so far in many sectors. Although their industrial usage date is recent, aluminum alloys have rapidly taken part in production [28].

Aluminum is used extensively in the automotive industry because it is a light metal, and its use is constantly increasing. Aluminum is used in the production of

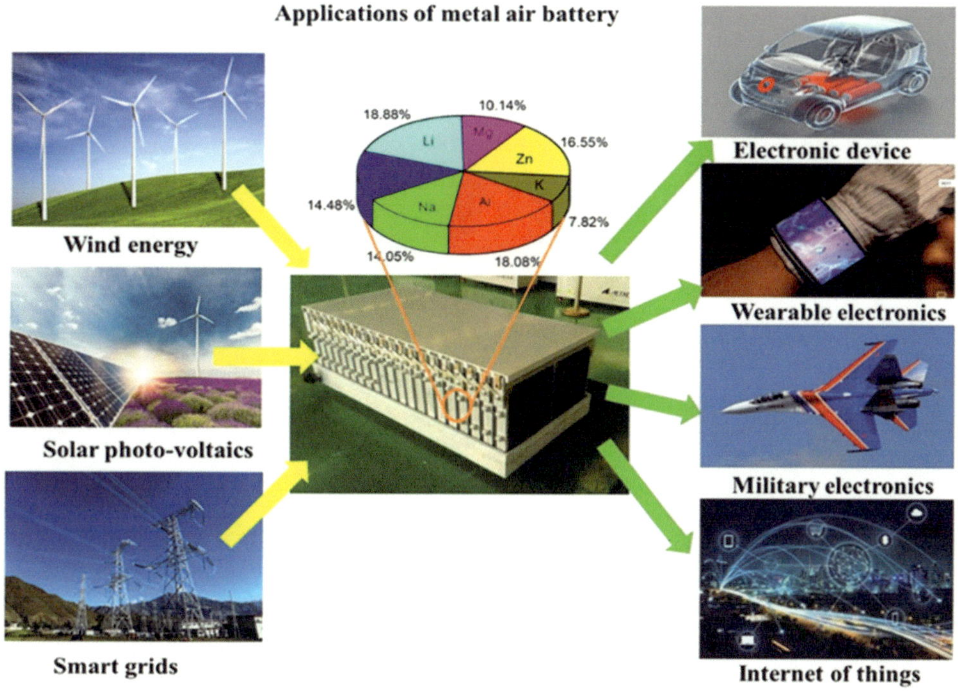

Figure 1.
Diagram of applications of aluminum batteries along with other different types of batteries [36].

radiators, engine parts, body sheets, and structural parts in the automobile industry. Aluminum is used in the construction of cargo transportation and passenger compartments in airplanes train transportation systems and in the production of ship hulls and propellers in the ship industry. Considering that energy will become more valuable in the future as a new area of use, aluminum batteries will find a wide range of applications (**Figure 1**). Aluminum-sulfur batteries are the first examples of these applications. With these batteries, it is possible to reach an efficiency of 250 Wh/kg. Another example is aluminum air-fuel cells [37].

Esthetic applications of aluminum alloys in the construction industry have a longer history than manufacturing and other strategic applications. In the construction industry, the needs could be met without the need for high technology, but according to the strength and corrosion properties of aluminum and in most applications, where both are required together, the aluminum industry had to carry out basic studies in technology and production methods, which resulted in the development of alloys and different production methods [38].

Author details

Emre Altaş[1], Shashanka Rajendrachari[2]* and Vutukuru Mahesh[3]

1 Department of Mechanical Engineering, Faculty of Engineering, Architecture and Design, Bartın University, Bartın, Turkey

2 Department of Metallurgical and Materials Engineering, Bartin University, Bartin, Turkey

3 Mechanical Engineering, SR University, Warangal, Telangana, India

*Address all correspondence to: shashankaic@gmail.com

References

[1] Santos MC, Machado AR, Sales WF, Barrozo MA, Ezugwu EO. Machining of aluminum alloys: A review. The International Journal of Advanced Manufacturing Technology. 2016;**86**:3067-3080

[2] Ma Z, Feng A, Chen D, Shen J. Recent advances in friction stir welding/processing of aluminum alloys: Microstructural evolution and mechanical properties. Critical Reviews in Solid State and Materials Sciences. 2018;**43**:269-333

[3] Abd El-Aty A, Xu Y, Guo X, Zhang S-H, Ma Y, Chen D. Strengthening mechanisms, deformation behavior, and anisotropic mechanical properties of Al-Li alloys: A review. Journal of Advanced Research. 2018;**10**:49-67

[4] StJohn D, Easton M, Qian M, Taylor J. Grain refinement of magnesium alloys: A review of recent research, theoretical developments, and their application. Metallurgical and Materials Transactions A. 2013;**44**:2935-2949

[5] Canakci A, Varol T. Microstructure and properties of AA7075/Al–SiC composites fabricated using powder metallurgy and hot pressing. Powder Technology. 2014;**268**:72-79

[6] Bhagat RB. Advanced aluminum powder metallurgy alloys and composites. In: ASM Handbook, Volume 7: Powder Metal Technologies and Applications. OH, USA: ASM Digital Library; 2013. pp. 840-858

[7] Liao J, Tan M-J. Mixing of carbon nanotubes (CNTs) and aluminum powder for powder metallurgy use. Powder Technology. 2011;**208**:42-48

[8] Vaidya M, Muralikrishna GM, Murty BS. High-entropy alloys by mechanical alloying: A review. Journal of Materials Research. 2019;**34**:664-686

[9] Suryanarayana C, Al-Aqeeli N. Mechanically alloyed nanocomposites. Progress in Materials Science. 2013;**58**:383-502

[10] Aboulkhair NT, Simonelli M, Parry L, Ashcroft I, Tuck C, Hague R. 3D printing of aluminium alloys: Additive manufacturing of aluminium alloys using selective laser melting. Progress in Materials Science. 2019;**106**:100578

[11] Parsons EM, Shaik SZ. Additive manufacturing of aluminum metal matrix composites: Mechanical alloying of composite powders and single track consolidation with laser powder bed fusion. Additive Manufacturing. 2022;**50**:102450

[12] Ozdemir F, Witharamage CS, Darwish AA, Okuyucu H, Gupta RK. Corrosion behavior of age hardening aluminum alloys produced by high-energy ball milling. Journal of Alloys and Compounds. 2022;**900**:163488

[13] Mahale RS, Vasanth S, Krishna H, Shashanka R, Sharath P, Sreekanth N. Electrochemical sensor applications of nanoparticle modified carbon paste electrodes to detect various neurotransmitters: A review. Applied Mechanics and Materials. 2022;**908**:69-88

[14] Pradeep N, Hegde MR, Rajendrachari S, Surendranathan A. Investigation of microstructure and mechanical properties of microwave consolidated TiMgSr alloy prepared by high energy ball milling. Powder Technology. 2022;**408**:117715

[15] Rajendrachari S. An overview of high-entropy alloys prepared by mechanical alloying followed by the characterization of their microstructure and various properties. Alloys. 2022;**1**:116-132

[16] Shashanka R, Uzun O, Chaira D. Synthesis of nano-structured duplex and ferritic stainless steel powders by dry milling and its comparison with wet milling. Archives of Metallurgy and Materials. 2020;**65**:5-14

[17] Shashanka R. Investigation of optical and thermal properties of CuO and ZnO nanoparticles prepared by *Crocus sativus* (Saffron) flower extract. Journal of the Iranian Chemical Society. 2021;**18**:415-427

[18] Shashanka R. Synthesis of nano-structured stainless steel powder by mechanical alloying-an overview. International Journal of Scientific & Engineering Research. 2017;**8**:588-594

[19] Nayak A, Shashanka R, Chaira D. Effect of nanosize yittria and tungsten addition to duplex stainless steel during high energy planetary milling. In: IOP Conference Series: Materials Science and Engineering. England: IOP Publishing; 2016. p. 012008

[20] Zhou S, Wu K, Yang G, Wu B, Qin L, Wu H, et al. Microstructure and mechanical properties of wire arc additively manufactured 205A high strength aluminum alloy: The comparison of as-deposited and T6 heat-treated samples. Materials Characterization. 2022;**189**:111990

[21] Rajendrachari S. Effect of sintering temperature on the pitting corrosion of ball milled duplex stainless steel by using linear sweep voltammetry. Analytical and Bioanalytical Electrochemistry. 2018;**10**:349-361

[22] Shashanka R, Chaira D. Optimization of milling parameters for the synthesis of nano-structured duplex and ferritic stainless steel powders by high energy planetary milling. Powder Technology. 2015;**278**:35-45

[23] Chaira D. Development of nano-structured duplex and ferritic stainless steels by pulverisette planetary milling followed by pressureless sintering. Materials Characterization. 2015;**99**:220-229

[24] Shashanka R. Fabrication of Nano-Structured Duplex and Ferritic Stainless Steel by Planetary Milling Followed by Consolidation. India: Educreation Publication; 2016. p. 146

[25] Rajendrachari S, Yilmaz VM, Karaoglanli AC, Uzun O. Investigation of activation energy and antibacterial activity of CuO nano-rods prepared by *Tilia tomentosa* (Ihlamur) leaves. Moroccan Journal of Chemistry. 2020;**8**:497-509

[26] Moghimian P, Poirié T, Habibnejad-Korayem M, Zavala JA, Kroeger J, Marion F, et al. Metal powders in additive manufacturing: A review on reusability and recyclability of common titanium, nickel and aluminum alloys. Additive Manufacturing. 2021;**43**:102017

[27] Farjana SH, Huda N, Mahmud MP. Impacts of aluminum production: A cradle to gate investigation using life-cycle assessment. Science of the Total Environment. 2019;**663**:958-970

[28] Wang H, Leung DY, Leung M, Ni M. A review on hydrogen production using aluminum and aluminum alloys. Renewable and Sustainable Energy Reviews. 2009;**13**:845-853

[29] Kaufman JG. Introduction to Aluminum Alloys and Tempers. USA: ASM international; 2000

[30] Toros S, Ozturk F, Kacar I. Review of warm forming of aluminum–magnesium alloys. Journal of Materials Processing Technology. 2008;**207**:1-12

[31] Knipling KE, Dunand DC, Seidman DN. Criteria for developing castable, creep-resistant aluminum-based alloys – A review. International Journal of Materials Research. 2022;**97**:246-265

[32] Min J, Xie F, Liu Y, Hou Z, Lu J, Lin J. Experimental study on cold forming process of 7075 aluminum alloy in W temper. CIRP Journal of Manufacturing Science and Technology. 2022;**37**:11-18

[33] Li G, Li X, Guo C, Zhou Y, Tan Q, Qu W, et al. Investigation into the effect of energy density on densification, surface roughness and loss of alloying elements of 7075 aluminium alloy processed by laser powder bed fusion. Optics & Laser Technology. 2022;**147**:107621

[34] Zhou B, Liu B, Zhang S. The advancement of 7xxx series aluminum alloys for aircraft structures: A review. Metals. 2021;**11**:718

[35] Bouali A, Serdechnova M, Blawert C, Tedim J, Ferreira M, Zheludkevich M. Layered double hydroxides (LDHs) as functional materials for the corrosion protection of aluminum alloys: A review. Applied Materials Today. 2020;**21**:100857

[36] Wang C, Yu Y, Niu J, Liu Y, Bridges D, Liu X, et al. Recent progress of metal–Air batteries—A mini review. Applied Sciences. 2019;**9**:2787

[37] Zhang J, Song B, Wei Q, Bourell D, Shi Y. A review of selective laser melting of aluminum alloys: Processing, microstructure, property and developing trends. Journal of Materials Science & Technology. 2019;**35**:270-284

[38] Georgantzia E, Gkantou M, Kamaris GS. Aluminium alloys as structural material: A review of research. Engineering Structures. 2021;**227**:111372

Chapter 2

A Review on Fundamentals of Grain Refining of Al-Si Cast Alloys

Ehab Samuel, Hicham Tahiri, Agnes M. Samuel, Victor Songmene and Fawzy H. Samuel

Abstract

Grain refining is considered one of the most important liquid metal processing processes for aluminum alloys. Three different types of grain morphology are possible: columnar, twin columnar and equiaxed. The present work reviews most of the theories that were proposed during the past three decades. These theories were mainly based on thermal analysis and thermodynamics to explain the mechanisms of grain refining of Al-Si based alloys, including the role of the master alloy used i.e., Al-B, Al-Ti, and Al-Ti-B alloys. Other aspects were also examined, mainly the interactions between Si and/or Sr and the grain refining master alloy, superheating of the molten metal as well as holding time prior to casting. This phenomenon is normally termed "poisoning" since it reduces the effectiveness of the added grain refiners. The effects of grain refining on the alloy microstructural characteristics, mechanical properties, machinability, hot tearing etc. have not been addressed in the present article.

Keywords: grain refining, master alloys, poisoning, theories, thermodynamics

1. Introduction

Grain refinement in Al-Si casting alloys is usually assessed by the presence of titanium Ti and boron B. Since the 1980s, thermal analysis has established itself as an important alternative for determining the degree of grain refinement and in predicting the degree of modification of the eutectic silicon. Grain refining is one of the most important liquid metal processing processes for aluminum alloys. Three different types of grain morphology are possible: columnar, twin columnar and equiaxed. It is well known that an equiaxed grain structure provides uniform mechanical properties, reduced hot tear, second phase distribution and microporosity on a fine scale [1–5].

Grain refining in aluminum alloys aims to increase the number of crystallization sites of the pro-eutectic phase (α-Al phase) and avoid columnar growth. In order to have a fine scale grain size, the most widely practiced way is to present effective nuclei in the liquid metal using the Al-Ti-B grain refiners which usually contain active seeds like $TiAl_3$, TiB_2, AlB_2 or $(Al,Ti)B_2$. Thermodynamic studies suggest that these latter particles convert to TiB_2, so that the titanium would diffuse into the $(Al,Ti)B_2$ particles while the aluminum diffuses out, resulting in the formation of TiB_2 [6–8].

 IntechOpen

While the Al-Si alloy system is widely used in industry, constituting around 85–90% of aluminum parts produced, the eutectic silicon in untreated alloys is often very coarse, leading to poor mechanical properties, mainly ductility. These properties are strongly influenced by the morphology of the eutectic silicon. The latter changes from its original raw structure of platelets to a less harmful and finer fibrous structure termed as eutectic Si modification which leads to a significant improvement in the mechanical properties of the products. Modification of the eutectic silicon is usually accomplished by adding certain modifying agents such as strontium Sr. However, over-modification can lead to the formation of porosities and returns the silicon to its original shape, again weakening the characteristics of the alloy. The addition of strontium in Al-Si alloys leads to a considerable increase in the amount of the α-Al dendritic phase and changes the shape of the dendrites. In the presence of a grain refiner like Al-Ti-B, the modifier reduction is considerable since the Sr-Ti interaction alternately changes the volume fraction of the dendritic phase of α-Al and the morphology of the silicon phase [9, 10].

This study aims to establish suitable grain refining mechanisms in Al-Si alloys to study the consequences of the refining-modification interaction in these alloys. Different time and temperature parameters on the thermal analysis curves are analyzed.

1.1 Solidification phenomenon

Grain morphology in as-cast alloys can be categorized as equiaxed or columnar. The grains of an equiaxed structure are nucleated in the liquid pool ahead of the solidification front on particles deliberately added as inoculants or present as impurities. In engineering alloys, the supercooling required to launch a grain onto an inoculant particle is produced by constitutional effects: the splitting of the solute between liquid and solid creates a solute profile ahead of the solidification front, lowering the local liquidus temperature. In the absence of equiaxed grains, growth is typically columnar; columnar grains grow approximately perpendicular to the direction of heat flow and have high elongation. The growth of columnar grains is blocked if there are enough equiaxed grains. Given the importance of grain morphology to the properties of a cast alloy, it is important to predict the conditions that cause the transition from columnar to equiaxed growth [11, 12].

During their growth, equiaxed grains can adopt various morphologies, depending on the rate of cooling and the content of solutes in the liquid metal and the degree to which a grain can grow before impact or collision with other grains. Equiaxed grains initially grow as spheroids, but a smooth solid-liquid interface becomes unstable as the radius of the grain increases. The grain becomes globular and then adopts a strongly branched dendritic structure. This evolution is accelerated by high cooling rates and dissolved solids content.

The morphology of the columnar grains is influenced by cooling carried out constitutionally ahead of the solidification front. In situations where supercooling is greater ahead of this front than at the front itself, the solid-liquid interface is unstable and results in dendritic structures. The greater the constitutional cooling, the more pointed and branched the dendrites.

In the absence of a constitutionally cooled region, the solid-liquid interface remains planar. The observed microstructure can be influenced by magnification of the dendrite arms after solidification is complete; the structure of a solidification front that has been quenched to preserve its structure is often markedly different from one formed without quenching under otherwise identical solidification conditions.

DOI: http://dx.doi.org/10.5772/intechopen.112987

1.2 Importance of thermal analysis

Thermal analysis could provide a reliable method for determining grain size, modifying the eutectic silicon and even quantifying the iron content of the alloy if the latter is equal to or greater than 0.6% by weight [13]. Thermal analysis consists in recording the evolution of the temperature of an alloy as a function of time during solidification. During cooling, heat is released and the temperature rises to a value close to the equilibrium value. This warming process is called recalescence. Supercooling, which represents a thermodynamic force, appears on the solidification curve as a drop in temperature below the equilibrium temperature of the reaction [14].

Grain refinement is a result of two separate processes: nucleation of new liquid aluminum crystals, followed by growth to a limited size. Both of these processes need a driving force which must be provided to the system through supercooling and supersaturation with respect to the equilibrium conditions of the real system. **Figure 1a** shows that during the entire first period of the solidification process, only those parts of the liquid metal which are in contact with the mold walls are cooled to such a degree that the nucleation of new aluminum grains can take place. **Figure 1b** shows that nucleation begins above the steady state growth temperature. This means that new crystals can be formed, not only at the first contact of the liquid bath with the cold walls of the mold, but also in the liquid.

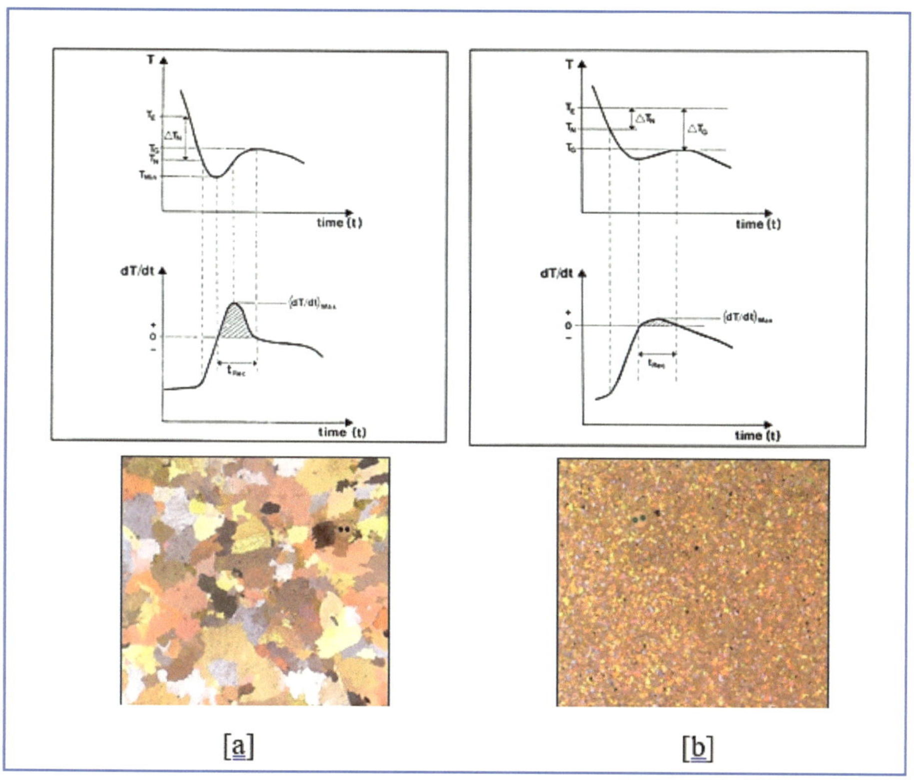

Figure 1.
(a) First part of a cooling curve and its derivative obtained from liquid metal close to the wall of the casting mold. (b) Cooling curve and its derivative of a sample to which titanium boride particles have been added. The nucleation temperature is below the liquid metal growth temperature. The recalescence phenomenon shows a very low value of $(dT/dt)_{max}$, indicating a sample whose grains are refined.

This phenomenon occurs due to particles with a high nucleation power (titanium borides). The latter becomes active at a supercooling of only 0.1–0.2°C when added to the liquid metal. Seed particles added to liquid metal must be effective substrates for heterogeneous nucleation in order to achieve grain refinement. However, nucleation can occur only if the liquid alloy is sufficiently cooled. In a solidification system, the remaining liquid pool can be cooled only if there is some solute in this liquid pool to limit solid growth or columnar competition with equiaxed solidification. Thus, efficient refining requires heterogeneous nucleation and growth restriction. Al-Ti-B refiners which are generally used for this purpose, consist of particles of TiB_2 with a diameter ranging from 0.1 to 10 μm and Al_3Ti particles with a diameter ranging from 20 to 50 μm, dispersed in an aluminum matrix [15]. Al_3Ti can be a very effective nucleant for aluminum, but this phase dissolves quickly when the refiner is added to commercial purity aluminum, because all the titanium content in the liquid metal is well within the limit of solubility [16, 17]. It is well accepted that some excess titanium (beyond that combined with B in TiB_2) is required for efficient nucleation [14].

Other temperature parameters seen in **Figure 1** correspond to:

T_E = The liquidus equilibrium temperature.

T_G = The steady state growth temperature of the molten metal.

T_N = The onset of nucleation temperature. TN is called the nucleation power of the particles present in the liquid metal. This point is most easily recognized by a sudden change in the derivative, as shown in the figure.

T_{MIN} = The temperature at which the newly nucleated crystals have grown to such an extent that the latent heat released swings out of equilibrium. After this time, the molten metal actually heats up to the steady state growth temperature. The period of time required for this heating is called the recalescence period (t_{Rec}).

1.3 Nucleation phenomenon

The grain refinement is carried out by the addition of master alloys of the Al-5Ti-1B and/or Al-Ti type in an embossed form. The addition rate of 1 kg/1000 kg gives Ti and B additions of 0.005% and 0.001%, respectively. Such an addition would typically produce grains of equiaxed structure with a grain size ranging from 100 to 150 μm in a small commercial pure aluminum ingot. The phenomenon of grain refining is directly linked to the process of nucleation and growth of aluminum grains. This is based on the nucleation ideas of Volmer and Weber [16].

The theory involves homogeneous and heterogeneous nucleation. In a solidified pure metal, the critical nucleus size for survival is given by:

$$r^*_{\text{homogène}} = \frac{-2\gamma_{sL}}{\Delta G_v} \tag{1}$$

The free energy barrier is given by:

$$\Delta G^*_{\text{homogène}} = \frac{16\pi\gamma^3_{sL}}{3\Delta G^2_v} \cong \frac{\Delta H_f \Delta T}{T_m} \tag{2}$$

where γ_{sL} is the surface energy of the interface separating the solid seed from the liquid in J/m^2.

By substituting $\Delta G_v = \frac{L_v \Delta T}{Ts}$ in Eq. (2), we obtain the following relation:

$$\Delta G^{*} = \left(\frac{16\pi\gamma_{sL}^{3}Ts}{3L_{v}^{2}}\right)\left(\frac{1}{\Delta T^{2}}\right) \tag{3}$$

where L_v is the latent heat of solidification per unit volume, ΔT is the supercooling (Ts − T) and Ts is the undercooling temperature. As for heterogeneous nucleation, the critical size of a nucleus is given by:

$$r^{*}_{h\acute{e}t\acute{e}rog\grave{e}ne} = \frac{-2\gamma_{sL}}{\Delta G_{v}} \tag{4}$$

Eqs. (1) and (4) are identical for both types of homogeneous and heterogeneous nucleation. The potential barrier that the germ must cross to reach its critical size is given by the following equation:

$$\Delta G^{*}_{h\acute{e}t\acute{e}rog\grave{e}ne} = \left(\frac{16\pi\gamma_{sL}^{3}}{3\Delta G_{v}^{2}}\right)f(\theta) \tag{5}$$

where f(θ) is a function of the contact angle θ on the substrate on which nucleation takes place. **Figure 2a** shows a nucleated solid on a substrate in a liquid. **Figure 2b** shows the variation of f(θ) with θ and since f(θ) is always ≤1, the critical free energy for heterogeneous nucleation is always less than or equal to that for homogeneous nucleation. However, it is clear that effective heterogeneous substrates are those with θ close to zero [10]. Undercooling ΔT values are of the order of 0.1–0.2°C for observable nucleation rates in commercial aluminum alloys with grain refiners. Therefore, clearly heterogeneous nucleation takes place. The simplified expression for the heterogeneous nucleation rate per unit volume in $m^3\ s^{-1}$ is:

$$I^{v}_{h\acute{e}t\acute{e}rog\grave{e}ne} = 10^{18}N_{v}^{p}\exp\left[\frac{-16\pi\gamma_{sL}^{3}f(\theta)}{3K_{B}\Delta S^{2}\Delta T^{2}}\right]$$

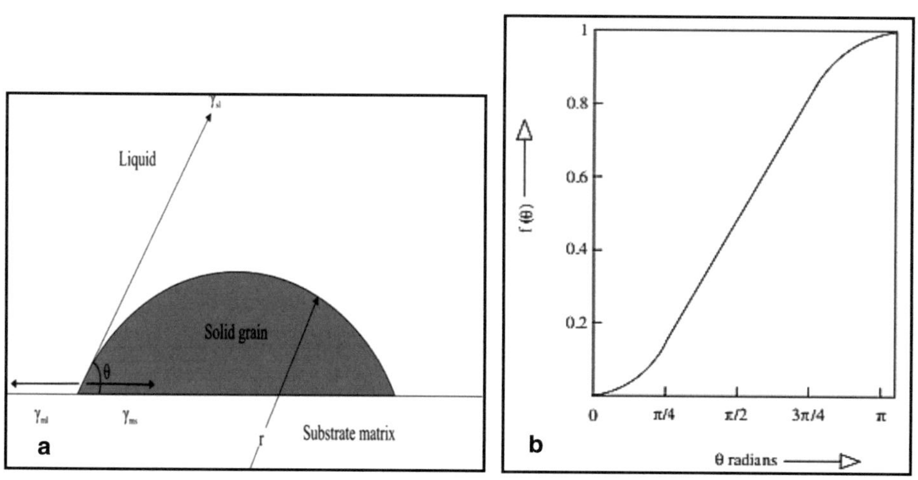

Figure 2.
Schematic representation showing (a) the formation of a spherical wetting of a solid S on a substrate, contact angle and surface tensions, (b) the variation of f (θ) with θ where f (θ) is equal to (2–3 cos θ + cos³θ)/4.

where K_B is the Boltzmann constant J/°C, N_v^p is the number of nucleant/m^3 and $I_{heterogène}^v$ is the heterogeneous nucleation rate of nucleant/m^3 s. Therefore, if the contact angle is near zero, the wetting of the substrate for nucleation is promoted and the rate of nucleation is increased.

When the nucleation sites are homogeneously dispersed in the liquid pool, the result is a fine grainy structure. The important topics for understanding the nucleation phenomena are summarized as follows: (1) the contact angle between the molten metal and the nucleation particles, (2) the interface energy between the molten metal, and (3) the nucleants and the coherence of the lattices of the nucleants and metal liquid. The presence of possible phases at different T in the liquid pool can be evaluated by comparing the free energy ΔG of the reactions.

Based on thermodynamic data, the calculated results are shown in **Figure 3**. It can be observed that $\Delta G(TiB_2)$ is much more negative than $\Delta G(Al_3Ti)$ and $\Delta G(AlB_2)$ in the range of T from 700 to 1200°K, while $\Delta G(Al_3Ti)$ is less negative than $\Delta G(AlB_2)$. In other words, the TiB_2 phase is easier to form than the Al_3Ti and AlB_2 phases. With increasing temperature, the changes of free Gibbs enthalpy of TiB_2 and AlB_2 are almost constant while that of Al_3Ti became small. In fact, the theoretical prediction indicates that Al_3Ti particles become unstable when the reaction temperature is increased. From the crystallographic point of view, and to further explain the high stability of the TiB_2 nucleation sites, the hybridization of the 3d orbital of titanium and the 2p orbital of boron is the main reason for the strong bond between these two elements. The bonding behavior between Ti and B layers is a combination of covalent and ionic nature.

1.4 Effect of overheating

As the Ti/B ratio is decisive for better grain refining, the casting temperature also plays a very important role when determining the grain size. Li et al. [18] studied the effect of overheating on pure aluminum before casting. **Figure 4** clearly shows the relationship between the superheat temperature and the average grain diameter. It is clear that as the temperature is increased from 725°C to 950°C, the average grain sizes

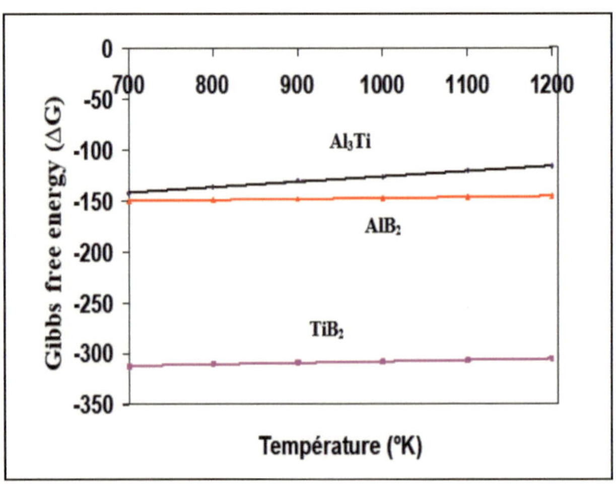

Figure 3.
Gibbs free energy of TiB_2, AlB_2, and Al_3Ti as a function of temperature.

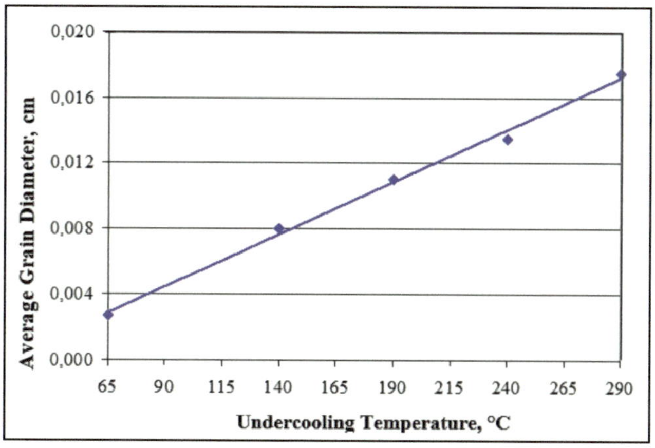

Figure 4.
Effect of undercooling temperature on the grain size of pure aluminum.

increase linearly. Casting temperature is a significant factor during the performance of grain refining. If the holding temperature is too high after inoculation, some fading or degradation occurs. This can be attributed to the growth and arrangement of TiB_2 particles, leaving a liquid alloy depleted of nucleating particles for efficient grain refining.

It is known that the size of the TiB_2 particles that form inside the aluminum depends on the temperature of the liquid metal; at high temperatures, the particles being formed are so large that they can settle to the bottom of the liquid bath by virtue of their greater density.

In general, overheating increases the grain size [11, 19], but in some cases it reduces it [20].

1.5 Master alloys

By adding three master alloys of the type Al-Ti-B, Al-Ti and Al-B with an excess of TiB_2 (Ti/B = 2.22), Lu et al. [21] examined the grain size in an Al-7%Si alloy (composition close to that of the hypoeutectic A356 alloy). The performance of these master alloys is shown in **Figure 5**. Indeed, the binary Al-Ti alloy is found to be less efficient, while the Al-B alloy is the strongest grain refiner in the Al-Ti alloys, since the grain size changes from 2000 μm to only 200 μm. From a certain level (0.1% by weight), the grain size remains constant even if the amount of master alloy is increased, hence the plateau obtained during supersaturation in master alloy.

In contrast, Al-B alloys show inefficient behavior in pure aluminum [22]. Similar observations were made by Sigworth and Guzowski [6] Cooper et al. [23] proved that the efficiency of residual nucleants of the TiB_2 and Al_3Ti type decreases with the number of recyclings of the Al-Si alloys, which explains the increase in grain size as a function of the number of repetitions of castings [24].

1.5.1 Al-Ti-B

The only controversial point in the ternary system of Al-Ti-B is related to the two borides TiB_2 and AlB_2. Both crystallize in the same crystal structure (hexagonal shape) with similar lattice parameters. The question is to determine if they form a single-

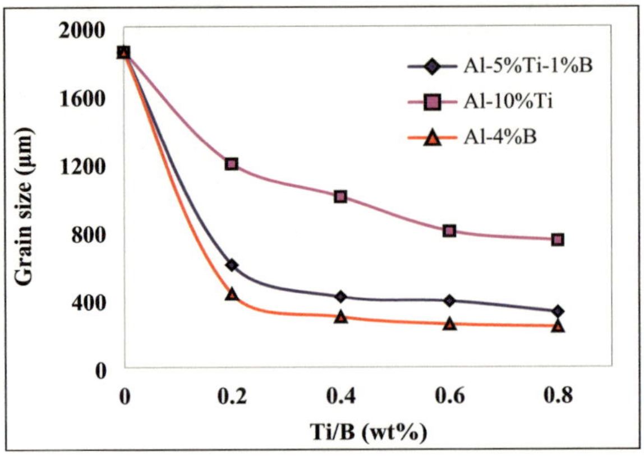

Figure 5.
Grain refining of 356 alloy using different master alloys.

phase continuous compound $(Al,Ti)B_2$ or if they coexist in a two-phase balance AlB_2+TiB_2. A single phase was previously assumed by Hayes et al. [25] considering all experimental data available at that time. Roger et al. [26] investigated a 1000°C isothermal section in the titanium-rich region using quantitative microprobe phase analysis of three molten ternary alloys. They found that TiB_2 is in equilibrium with all Al-rich Al-Ti binary phases. The aluminum-rich corner of the ternary phase diagram was first calculated by Hayes et al. [27]. No ternary parameter was used in this work.

The question whether TiB_2 and AlB_2 exist in two separate phases or as a solid continuous solution has not been resolved. Zupanic et al. [28] investigated arc melted alloys in the triangular composition Al-AlB_2-TiB_2 and found four solid phases: (Al), AlB_{12}, AlB_2 and TiB_2. The AlB_{12} phase, which is stable at very high temperatures in the Al-B binary system, decomposes during annealing below 900°C. Both borides AlB_2 and TiB_2 were found to coexist even after 1000 h at 800°C. Formation of mixed diboride $(Al,Ti)B_2$ was not observed [29]. Fjellstedt et al. [30] have produced Al-rich alloys by several different sample fabrication methods. They concluded that only the maximum solubility of aluminum in TiB_2 (up to 0.15 wt% Al) and titanium in AlB_2 (up to 0.2 wt% Ti) exists at 800°C, hence a continuous phase $(Al,Ti)B_2$ is not stable.

Based on the entity of this data, the AlB_2 and TiB_2 phases were modeled in the work of Gröbner et al. [31] as two separate phases without any solubility. No ternary phase or ternary solubility exists in this system, no ternary parameter is needed for the calculation. The complete phase diagram for the Al-Ti-B system can be calculated from the binary data and the extrapolation. An isothermal section at 500°C is given in **Figure 6**. It is important to note that the phase equilibria in the Al-rich corner are not changed significantly if a homogeneous ideal range of $(Al,Ti)B_2$ solution has not been assumed. The reason for this is the much higher thermodynamic stability of TiB_2 compared to that of AlB_2.

When an Al-Ti-B type master alloy is added to an alloy such as A356 (~7%Si), several intermetallic phases are created. Among these, there is mention made of intermetallics of the $(Al,Si)_3Ti$ [32, 33] and Ti_6Si_2B [31] type. Ramos et al. [34] carried out detailed research on the ternary phase Ti_6Si_2B which belongs to the Ti-Si-B system. Thanks to X-rays, this phase is characterized by a hexagonal crystalline structure with lattice parameters a = 0.68015 nm and c = 0.33377 nm, and it forms a liquid through the following peritectic reaction: L + TiB + Ti_5Si_3 ↔ Ti_6Si_2B.

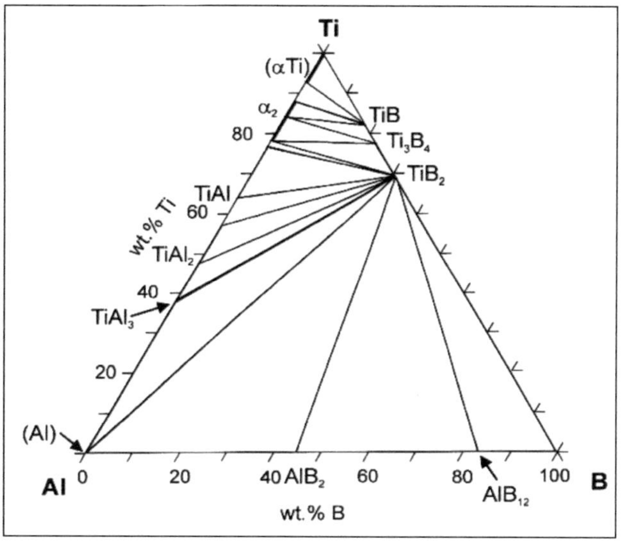

Figure 6.
Calculated isothermal section of the Al-Ti-B ternary system at 500°C.

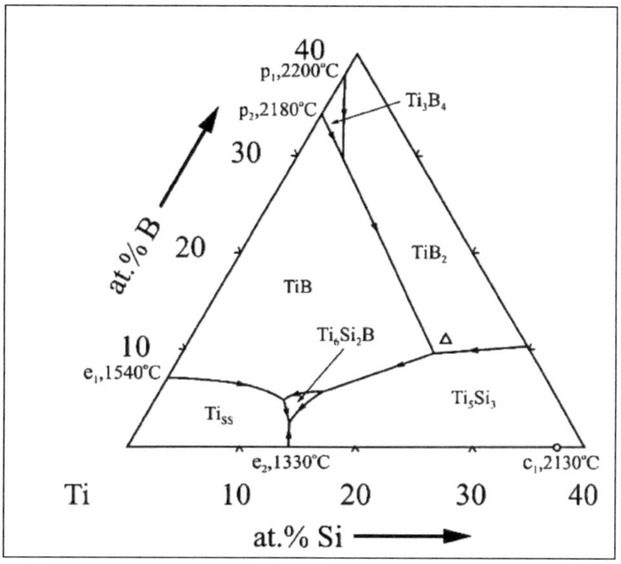

Figure 7.
Projection of the liquidus of the Ti-Si-B system in the Ti-rich part. The symbol Δ marks the composition of the Ti_6Si_2B phase.

As shown in **Figure 7**, which presents a liquidus projection of the Ti-rich part of the Ti-Si-B system, five regions of primary solidification coexist, namely, Ti_{ss}, Ti_5Si_3, Ti_6Si_2B, TiB and TiB_2. At 1200°C, the Ti_6Si_2B phase is formed in two-phase fields: Ti_{ss}, Ti_5Si_3 and TiB. Solidification ends with an invariant ternary eutectic Ti_{ss} + Ti_6Si_2B + Ti5Si3, which should correspond to the lowest liquidus temperature in the region shown in the figure. Based on the invariant reaction temperatures in the Ti-rich side of the Ti-Si [35] and Ti-B [36] systems, this ternary eutectic temperature should be less than 1330°C.

1.5.2 Al-Ti

One of the effects of the presence of titanium Ti in an aluminum alloy is the reduction in grain size. However, this reduction is no longer achievable in pure aluminum since the number of grains per unit length increases, and consequently, the size of the grains also increases as shown in **Figure 8**. This size becomes almost constant when the content in titanium is between 0.08% and 0.13% before it begins to increase again with increasing percentage of titanium [36].

Grain refinement in aluminum alloys by the addition of Al-Ti master alloys has been widely applied and studied in recent years. The grain refining mechanism by Al-Ti is not much doubted and can be explained by the action of Al_3Ti particles as heterogeneous nucleating centers [37] as well as by the peritectic theory. The size, morphology and quantity of the nuclei of the different microstructures of the Al-Ti master alloy seem to be important factors in determining the degree of grain refinement. The efficiency of the grain refiner depends on its chemical composition and its processing parameters such as the maintenance at such a temperature, the contact time, the mechanical agitation and the rate of cooling. To show the effect of increasing the percentage of titanium, Simensen [38] investigated a series of Al-7%Si alloys with a cooling rate of the order of 1°C/s. Al-10%Ti rods were added to the liquid metal making alloys with Ti in the range of 0.01–0.18%Ti. The cooling curves showed that at first the grains began to grow at some supercooling. The gradual addition of titanium increased the growth temperature of the alloys according to the following equation: T_{growth} = 613.2°C + 30.2%Ti (by weight). As for the grain size of the alloys, it was reduced from about 2000 μm to 250 μm when the titanium content increased from 0.01% to 0.12%. The best results were obtained when the $Al_3(Ti,Si)$ phases were nucleated on the TiB_2 particles during cooling, whereas the aluminum grains which form on the $Al_3(Ti,Si)$ intermetallics yield the fine-grained material.

On the other hand, Li et al. [39] investigated the effect of various microstructures on grain refinement by Al-Ti master alloys synthesized at high temperature by mixing aluminum with titanium. The results of their work showed that the variation in experimental parameters, such as the stoichiometric ratio of the initial powders, the particle sizes of the powders, the use of fluxes, etc., led to the formation of various structures of the master alloys, in particular sizes, morphologies and quantities of

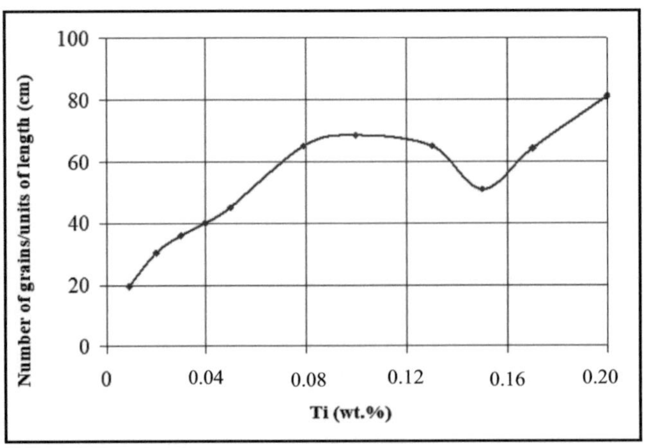

Figure 8.
Number of grains per unit length as a function of Ti content in pure aluminum.

Al_3Ti particles. The intermetallic particles showed needle-like morphology in the lack of aluminum powder in the initial mixture which was related to the higher reaction temperature (**Figure 9a**). The same particles had needle-like and blocky forms at the same time (mixed morphology) which resulted from the lower reaction temperature (**Figure 9b**). Block-like Al_3Ti crystals were formed at the lower reaction temperature when excess aluminum powder was added to the initial mixture (**Figure 9c**).

In order to clearly highlight the effect of the increase in titanium on the grain size and on the growth temperature Tc, Tøndel and Arnberg [40] studied the behavior of a binary alloy of the Al-10% Si type, with Ti additions through an Al-6%Ti master alloy. Two series of alloys were prepared: series A cast in a cold mold (0.02–1.5% Ti) and series B cast in a preheated mold (0.02–0.2% Ti). It was essential to compensate for the effect of differences in silicon content on the liquidus temperature before studying the effect of titanium additions on the recorded Tc growth temperature. Temperature data therefore compensated for the true deviation in Si content from a chosen low composition, 9.6%Si, in the claim that Ti additions do not change the slope of the liquidus line of the Al-Si system in small temperature intervals. The calculation was made with a polynomial that describes the liquidus temperature as a function of the Si content in hypoeutectic Al-Si binary alloys: T_{liq} (°C) = 660–5.59Si–0.14Si.

The phase identification clearly shows that when the Ti content is increased, intermetallic particles will appear in the solidified material. An example of a large Ti $(Al,Si)_3$ type crystal (~250 μm) is found in a B-series sample. Microprobe analyses of different samples show that Al_3Ti arises when silicon varies from 12 to 13% in the solution. This may, in fact, be an indication that the Al_3Ti particles of the master alloy have survived in the Al-Si liquid metal which thereby reaches equilibrium composition. The illustration in **Figure 10** suggests that the peritectic composition at 0.15% Ti is shifted to lower Ti concentrations when silicon is present because the particle is too large to be a result of exceeding the 0.15% Ti limit by only 0.023% Ti [41–43].

1.5.3 Al-B

The use of Al-B type master alloys (with 1–4%B) to achieve grain refining in Al-Si alloys is very common since this type of refiner is the most powerful in this type of alloys [43, 44]. The main particles that favor germination sites are AlB_2 (more stable in Al-Si) and AlB_{12}. The average AlB_2 particle size varies with the B content in the master alloy as shown in **Figure 11a**. It is obvious that the small additions in B have a remarkable effect on the solidification process of the alloy. **Figure 11b** shows the

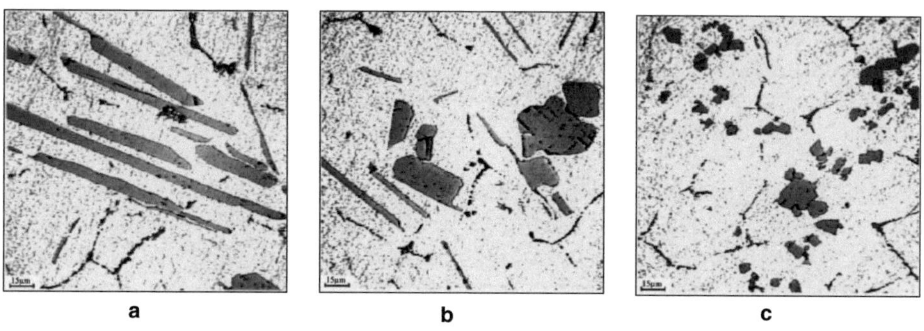

Figure 9.
Microstructures Al-Ti master alloys. (a) Master alloy with acicular Al_3Ti particles. (b) Master alloy with a mixture of acicular and blocky Al_3Tiparticles. (c) Master alloy with blocky Al_3Ti particles.

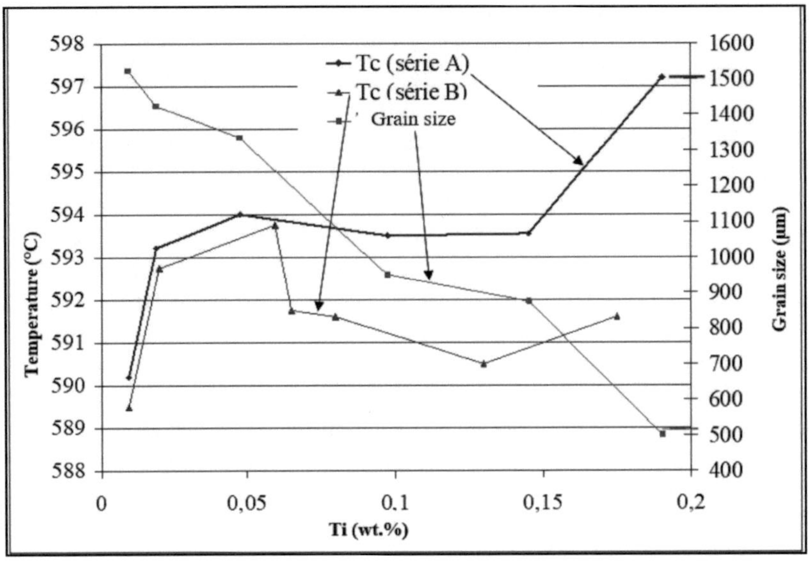

Figure 10.
Growth temperature Tc and grain size as a function of Ti.

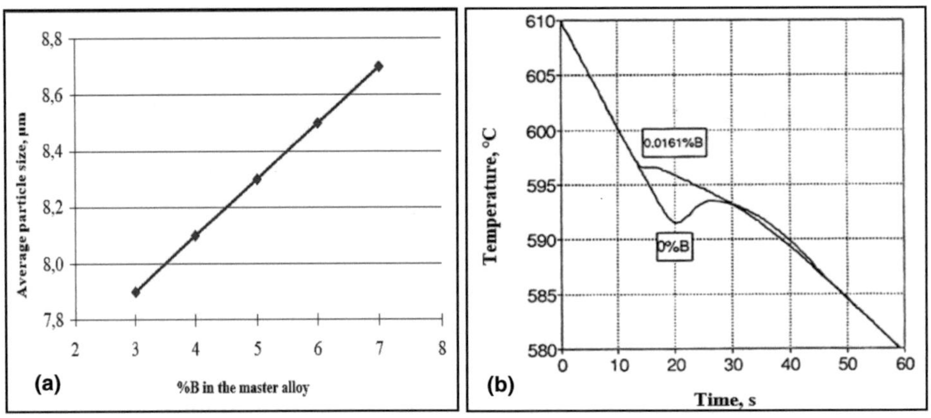

Figure 11.
(a) Average particle size of AlB_2 as a function of the content of B in Al-B, (b) cooling curve of the Al-9.6%Si alloy as a function of the added B.

effect of boron additions on the first part of solidification of an Al-9.6%Si alloy. The cooling curve increased by ~2 to 3°C when 161 ppm B was added. The addition of B eliminates the phenomenon of supercooling and recalescence on the solidification curve.

Several concepts of the grain refinement mechanism of B on hypoeutectic Al-Si alloys have been adopted: the effect of B grain refinement on the α-Al phase and on the eutectic silicon with different additions of master alloys at 850°C was studied by Wang and Bian [45], where master alloys formed under different temperature conditions were studied to explore the morphologies of the AlB_2 particles; the sample slowly cooled with the addition of grain refiner was made to explore the mechanism of refinement. The master alloy can refine not only the α-Al dendritic phase, but the eutectic silicon. Theoretical analysis indicates that although the AlB_2 particles do not participate directly in the pure Al nucleation process in the presence of silicon, they provide a substrate for the precipitation of a small silicon content on which α-Al will grow without supercooling. As the temperature decreases to the eutectic line, AlB_2 later nucleates the eutectic silicon; AlB_2 particles appear in two different morphologies, namely hexagonal platelet and tetradihedron morphology which depend on temperature processing conditions.

As grain refining is an important process in industrial practice and has been the subject of much study, microstructural characterization of master alloys is useful in monitoring their production to ensure consistency of performance [43, 46]. AlB_2 particles appear in master alloys in two different morphologies, namely, as hexagonal platelets or as a regular tetradihedron. This difference in morphology depends on the state of treatment. The hexagonal platelet morphology is favored by low temperature production, the reaction between Al and KBF_4 is inactive, and the solubility of B in liquid aluminum is rather low. The formation of the AlB_2 phase is affected by the long distance diffusion of B atoms, which makes the obvious growth trend of the crystal. The growth of the AlB_2 particle proceeds by the diffusion of B atoms along <1120> at the edge of the crystal platform, the diffusion rate of B along <0001> is negligible. The schematic representation is shown in **Figure 12**.

The tetradihedron morphology is tilted to be formed at high temperature. The chemical reaction of Al and KBF_4 proceeds rapidly at high temperature, therefore, the content of B in the aluminum liquid bath is higher at high temperature than that at low temperature. The diffusion of B atoms has relatively little influence on crystal growth, each plane of AlB_2 grows with almost the same speed. The diffusion velocity of B along <0001> cannot be neglected. When the diffusion velocity of B along < 1120> is a little higher than that along <0001>, the tetradihedron morphology will be formed as shown in **Figure 12b**. No evidence that the morphology has an influence on the efficiency of the refiner was found, however, it seems that the hexagonal plate is always located in the center of two silicon flakes. The tetradihedron morphology is tilted to lie at the center of α-Al [43].

Very small additions of boron to Al-Si alloys lead to the precipitation of aluminum borides. To the Al-Si eutectic liquid, an addition of about 0.01% by weight of B is sufficient for this effect. It is the dominant feature of the calculated Al-B-Si phase diagram section at 0.1%B by weight, as shown in **Figure 13**. It looks like the binary Al-Si diagram with just AlB_2 as the additional equilibrium phase. The other boride,

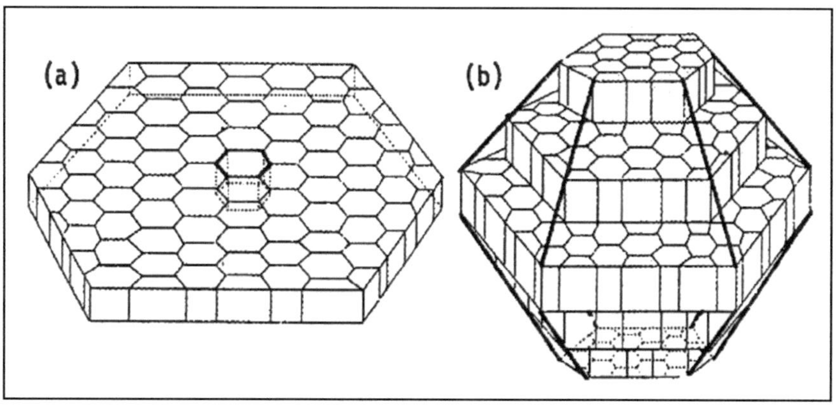

Figure 12.
Schematic representation of AlB_2 morphologies; (a) hexagonal insert, (b) tetra-dihedron.

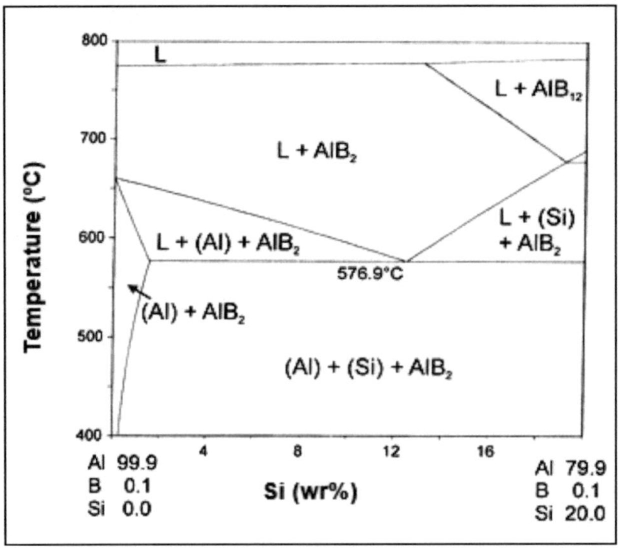

Figure 13.
Section of the Al-B-Si system calculated at 0.1%B. The composition of the ternary liquid eutectic is 12%Si and only 0.01%B; the temperature is 0.1°C below 577°C of the Al-Si binary eutectic system.

AlB_{12}, precipitates out of the liquid at high temperature and high in Si. In order to completely dissolve 0.1%B by weight, the liquid must be heated above 775°C. Using the Si-B master alloy, all boride particles should be formed in-situ during solidification, presumably in fine distribution and acting as strong nucleation sites. Only the binary peritectic reaction in the reverse direction, $L + AlB_{12} \rightarrow AlB_2$ is observed upon superheating at $T = 972 \pm 5$°C [41–47]. If AlB_2 formation is also suppressed by cooling Al-Si-B liquid alloys, the metastable phase diagram should be considered. Such a calculation proves that the relevant phase boundaries in **Figure 14** are virtually unchanged, just the saturation phase AlB_2 is replaced by AlB_{12}. This minor difference is also demonstrated by the calculated ternary eutectic:

$$\text{L Stable} = (\text{Al}) + (\text{Si}) + \text{AlB}_2 \text{ at 576.9°C and L with 12.5\%Si, 0.010\%B (wt.\%)},$$

$$\text{L Metastable} = (\text{Al}) + (\text{Si}) + \text{AlB}_{12} \text{ at 576.89°C and L with 12.5\%Si, 0.011\%B (wt.\%)}.$$

Both equilibria are just slightly below the calculated binary eutectic L = (Al) + (Si) at 577°C and L with 12.5% Si by weight. Due to the slow formation of AlB_2 in the liquid metal, it is very likely that the next stable phase, AlB_{12}, will be formed instead. The particles found in the center of the grains are in fact B-rich, but it can be difficult to distinguish between AlB_2 and AlB_{12} by backscattered electron and X-ray mapping techniques [48, 49]. The work carried out by Gröbner et al. [31] proves that, from the ternary phase diagram and thermodynamics, the two Al borides could be equally well formed. The fact that boron is a very efficient grain refiner in Al-Si alloys, but not in pure aluminum [4], is convincingly explained by the additional presence of dissolved silicon with a high growth restriction factor [50].

Figure 14 gives an overview of the grain size when the A356 alloy is treated by three grain refiners. In the absence of any addition, the grain size amounts to about 1850 μm in the base alloy, see **Figure 14a**. After an addition of 0.08%Ti using the Al-10%Ti master alloy, the size drops to approximately 800 μm, see **Figure 14b**.

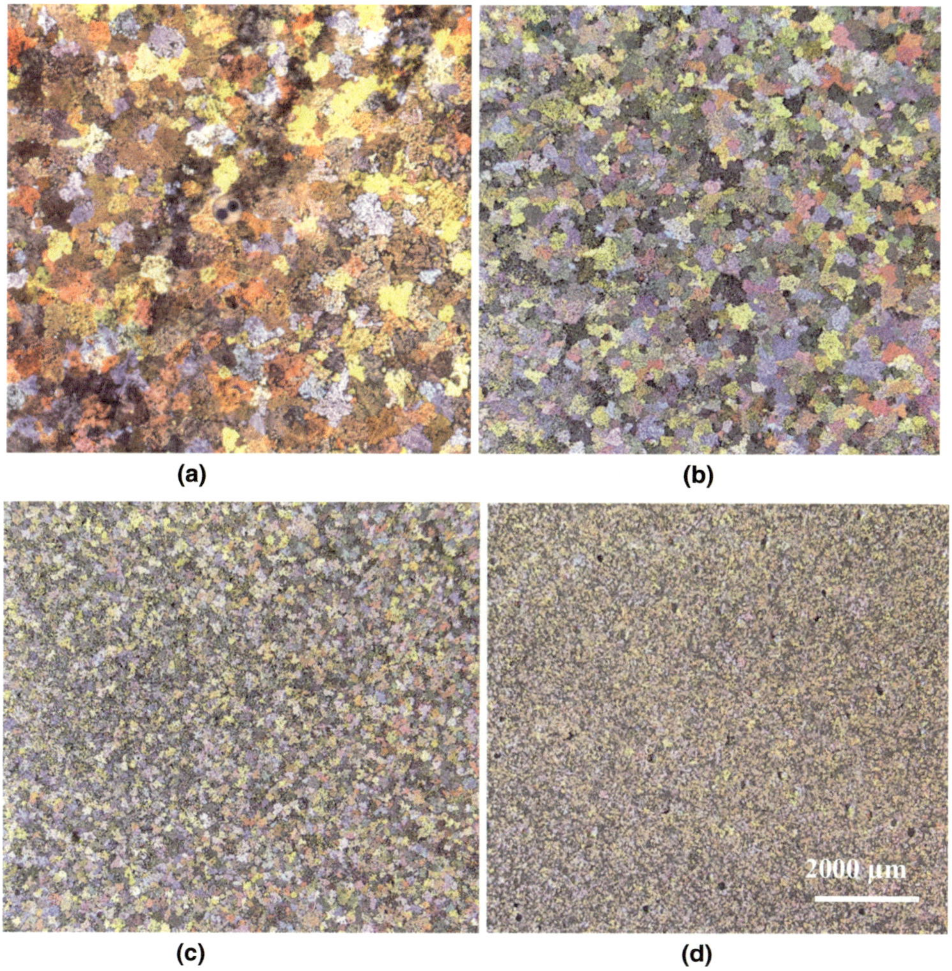

Figure 14.
Evolution of the grain size of the A356 alloy: (a) no addition, (b) Al-10%Ti, (c) Al-5%Ti-1%B, (d) Al-4%B.

The aluminum grain size continues to decrease when 0.08% Ti is added using Al-5% Ti-1%B, see **Figure 14c**. A minimum value is obtained thanks to the boron-based refiner (Al-4%B) since the grain size is reduced to 200 µm, see **Figure 14d**. Adding excess titanium or boron has no effect on reducing the aluminum grain size. On the contrary, an overdose of titanium or boron can lead to deleterious effects on the microstructure of the alloy, and consequently on its mechanical properties.

1.5.4 Effect of master alloys

The literature concerning the influence of the addition of Al-Ti and Al-B master alloys on grain refining is very voluminous. Li et al. [51] investigated the effect of the Ti:B ratio on the solidification structure of a molten aluminum arc which is similar to the welding arc process. Al blocks were prepared, a hole with a diameter of 3.5 mm and another with a diameter of 7.5 mm deep were drilled in the center of each aluminum block in order to hold the two types of powder used (Al-Ti and Al-B). Grain size measurements were made using the linear intercept method, and were conducted at the edge, middle and center positions of each weld. The results are plotted in conventional form in **Figure 15**. This plot shows a minimum grain size at about 0.07% Ti.

Such a minimum has never been reported previously in welding or casting. However, one must be careful when interpreting such graphs because the Ti:B ratio also affects the performance of grain refiners.

The benchmarks shown in **Figure 15** are obtained for different Ti:B ratios. The minimum occurs when the Ti:B ratio approaches an atomic ratio of 2:1, which is the stoichiometric ratio for the formation of TiB_2 and the highest mole number of TiB_2 in the solder. Thus, it is concluded from these data that a stoichiometric TiB_2 is the most effective compound for a good grain refinement of aluminum under these experimental conditions. This disagrees with a few previous reports [52] which point out that excessive Ti is necessary for good grain refining under casting conditions.

1.6 Theories of grain refining

Grain refining is an important technique for improving the properties of aluminum products. Various explanations have been presented in order to provide, a suitable mechanism for grain refining such as nucleating particle theories and phase diagram theories. Both categories of the theories are about the two types of particles present in Al-Ti-B master alloys. Particle theories or boride theory suggest that nucleation occurs on the borides in the master alloy (TiB_2, AlB_2 and $(Ti,Al)B_2$), while phase diagram theories explain grain refinement by nucleation on the $TiAl_3$ properitectic phase.

1.6.1 Particle nucleation

Cibula [53] proposed that nucleation is produced on borides or carbides when the latter are present. Borides are added by a master alloy, whereas carbides are formed by a reaction of residual carbon present in the liquid metal with additional titanium which results in a TiC-like form. The nucleation behavior of all borides can be

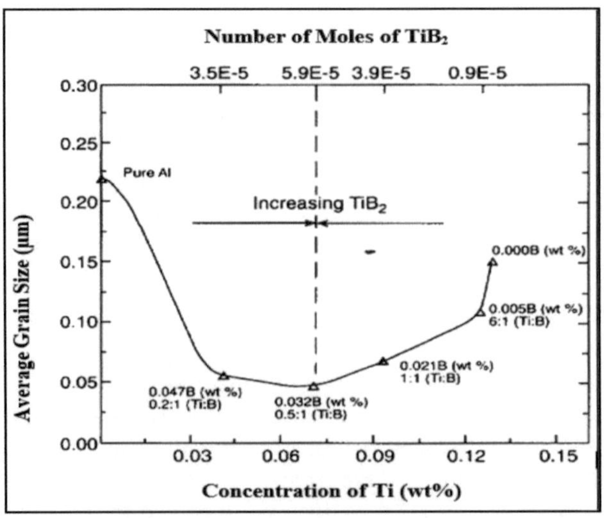

Figure 15.
Effect of of Ti and B additions on the average grain size.

discussed concurrently, since TiB_2 and AlB_2 are known to be isomorphic and hexagonal, with lattice parameters changing only slightly, having a = 0.30311 nm and c = 0.32291 nm, and a = 0.3009 nm and c = 0.3262 nm, respectively. The boron mixed phase is formed by replacing titanium atoms in the lattice with aluminum atoms. The stability of the $(Al,Ti)B_2$ phase is not known; however, it is thought to convert to TiB_2 after a long hold time. **Figure 16** shows the crystal structure of these two locations [54, 55].

When the Al-Ti-B master alloy is added, the titanium is present in hypoperitectic amounts (less than 0.15% Ti, **Figure 16**), where boron particles are often found at the grain centers, with titanium-enriched dendrites growing outside of them. This evidence suggests that borides nucleate in the α-Al phase. However, for other reasons, borides were thought to be less efficient nucleation sites than Al_3Ti. In the master alloys, the borides are pushed or rejected towards the grain boundaries while the aluminides are at the centers of the grains [21]. Recently, Schumacher and Greer [55] confirmed that borides are pushed to grain boundaries and no grain enhancement is observed when there is no dissolved titanium. Additionally, borides are known to need some supercooling to nucleate aluminum, while aluminides need none.

Al_3Ti particles are known to be strong nucleating bodies. If titanium is present at hypo-peritectic concentrations, a dramatic grain enhancement is observed. Al_3Ti are found in the center of the grains at concentrations where they are stable, and multiple orientation ratios have been recognized between Al_3Ti and the aluminum matrix [49]. Obviously, it can be concluded that Al_3Ti is a better nucleant than TiB_2. Phase diagram theories have been developed to explain how Al_3Ti could be an active nucleant with hypo-peritectic compositions.

1.6.2 Theory of phase diagrams

The theories in this category are grouped under this heading because each theory suggests that the grain refinement is caused by a peritectic reaction on the primary Al_3Ti particles. In general, it has been suggested that a peritectic point shifts at low titanium concentrations (e.g. 0.05% Ti) caused by the addition of boron, and that this is the reason for the grain enhancement [54]. Therefore, it has generally been assumed that there is a peritectic Al-Ti-B ternary, and speculations and theories have been based on this assumption. The first attempt to explain the grain refining mechanism dates back to 1951. Crossley and Mondolfo [49] proposed a peritectic theory based on the peritectic reaction in the phase diagram of the Al-Ti system as:

$$\text{Liquid} + TiAl_3 = \alpha - Al\ (\text{solid solution}).$$

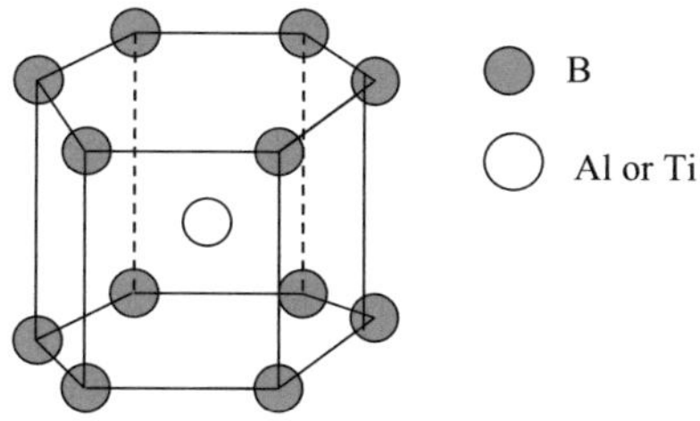

Figure 16.
Crystalline structure of AlB_2 and TiB_2.

It is reasonably clear that the titanium aluminum crystals added by the master alloy are active nucleants and that the observed fading is due to the dissolution of these nucleants with time. Davies et al. [56], and Maxwell and Hellawell [57] observed $TiAl_3$ particles at the center of the grains of the α-Al dendritic phase. The cooling curves published by Arnberg et al. [58] also support the order of nucleation, i.e., they show no supercooling but a nucleation temperature (T_n) above the melting point (T_f) of the liquid metal. This observation implies that nucleation occurs by a peritectic reaction around the peritectic temperature (665°C) which is higher than the melting point of pure aluminum.

Although the peritectic theory successfully explains the behavior of Al-Ti type master alloys, no consensus has emerged to explain the increased efficiency of commercial grain refiners containing titanium Ti and boron B. The authors of the peritectic theory suggest that the improved performance of boron is due to the shift of the peritectic composition from 0.15% Ti towards the aluminum end of the phase diagram, which ensures the thermodynamic stability of $TiAl_3$ at low levels of addition of Ti (~0.02%). Determined phase relationships show that the solubility of $TiAl_3$ is practically unaffected by the presence of boron B. However, contrary to these thermodynamic predictions, Mondolfo et al. [48] obtained experimental data indicating the effect of boron by shifting the peritectic to the Al-rich end of grain seems to be far from being achieved. **Figure 17** presents the aluminum-rich part of the Al-Ti system diagram which clearly shows the domain of existence of the different phases with their formation temperatures.

1.6.3 Theory of peritectic transformation

This theory was very popular in the late 1980s and early 1990s supported by Vader and Noordegraaf [59], and Bäckerud et al. [60, 61]. This theory assumes that $TiAl_3$ is a stronger nucleant than TiB_2. Therefore, it tries to explain how borides could slow down the dissolution rate of $TiAl_3$ when an Al-Ti-B master alloy is added to the liquid aluminum bath, so that the more powerful nuclei remain active longer. It suggests that the borides form a shell around the aluminides and therefore slow down the

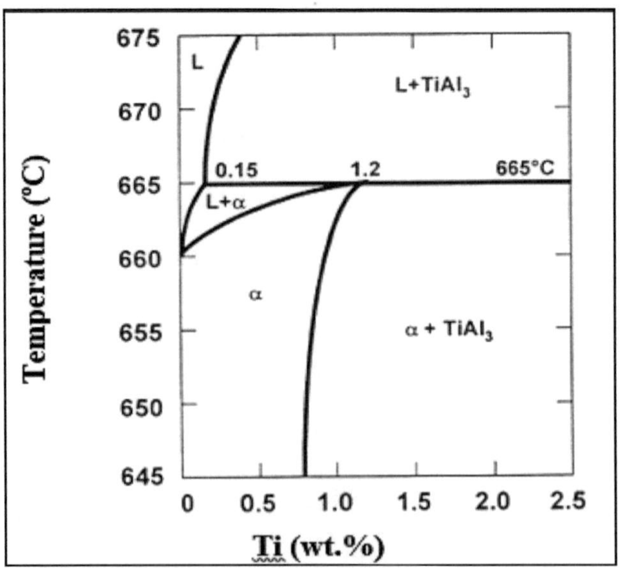

Figure 17.
Al-Ti binary diagram (Al rich corner).

dissolution of the aluminides as the diffusion must proceed through the boride shell. The finally dissolved aluminide leaves a liquid cell inside the boride shell, approximately peritectic in composition. Peritectic reaction takes place to form α-aluminum and growth occurs from there. A schematic representation of the peritectic reaction is shown in **Figure 18**.

Although this theory appears to fit experimental results, there is strong evidence against it, particularly that described by Johnsson et al. [62]. Borides are very stable in liquid aluminum alloys, compared to $TiAl_3$ particles, to hypo-peritectic titanium compositions ($\sim$ 0.15% Ti by weight). The peritectic theory suggests that borides are more soluble than $TiAl_3$, because the borides must dissolve in the liquid bath so that they can reprecipitate onto the more slowly dissolved $TiAl_3$ particles in the titanium-rich region produced by its dissolution - which does not seem possible. Even with boron in the master alloy, $TiAl_3$ still dissolves after a few minutes at high temperatures.

Johnsson [63] melted and re-solidified a hypo-peritectic alloy and found that the efficiency of grain refining did not change with the number of cycles. If the peritectic mechanism were to occur, one would expect the efficiency of grain refining to decrease with the number of repetitions, as this would further allow the diffusion of titanium; therefore, the peritectic reaction would cease to occur. Boride shells have been found in the grains of the aluminum, although it is inconclusive that these were the site of nucleation. If they acted as nucleants, this was not the dominant mechanism, as more often boride particles were found in the grain center at hypoperitectic concentrations of titanium. Therefore, the evidence suggests that the peritectic mechanism did not work.

1.6.4 Theory of hyper-nucleation

Jones and Pearson [5] assumed the concept of hyper-nucleation theory at the TiB_2/liquid metal interface. They proved that when titanium is in excess, the titanium atoms segregate at the TiB_2/liquid metal interface, providing a stabilized layer of atoms on the surface of the TiB_2 crystals. This layer, being a full solution of Ti and Al, was predicted to remain stable above the melting point of pure aluminum, i.e., it exists in the liquid metal before casting. During cooling, such a layer will allow the growth of the primary α-Al phase without any supercooling. Although this concept seems the most promising, no experimental evidence is there to support it. **Figure 19** presents a model of the hyper-nucleation theory [5, 64–66]:

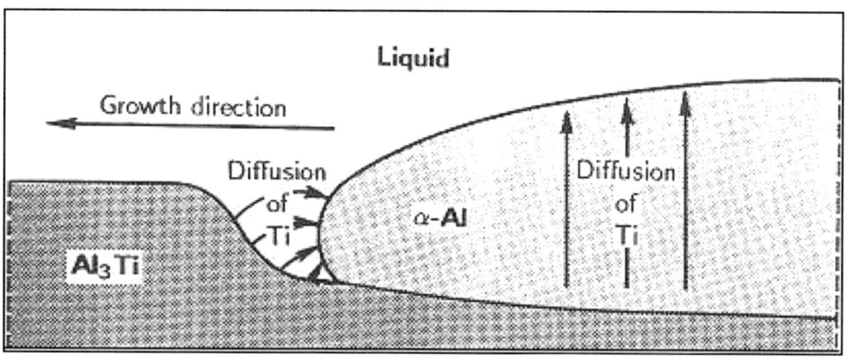

Figure 18.
Schematic presentation of peritectic reaction.

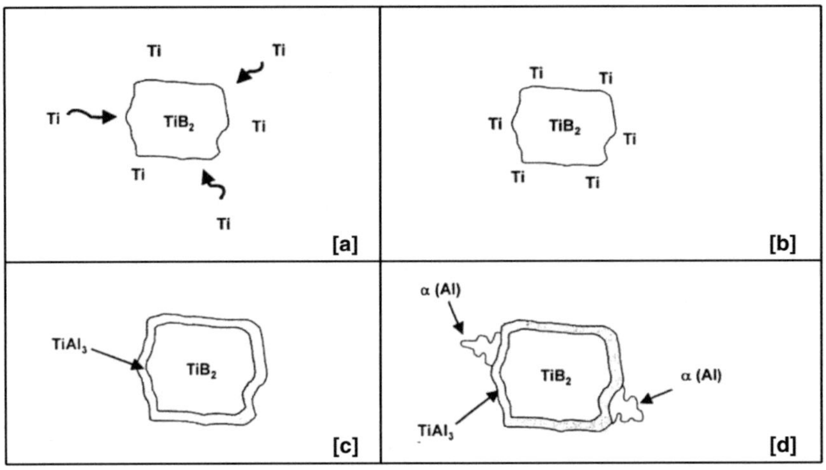

Figure 19.
Schematic presentation of the theory of hypernucleation. (a) Excess of Ti (Ti/B > 2.21) in solution, (b) Ti segregation at the interfaces TiB_2-liquid metal, (c) formation of layers of $TiAl_3$ on TiB_2, (d) nucleation of α-Al through peritectic reaction.

1.6.5 Duplex theory of nucleation

Mohanty et al. [67–70] suggest that the formation of Al_3Ti is caused by a concentration gradient of titanium towards the boride particles, constituted by an activity gradient towards the borides. Due to the local equilibrium near the borides, the Al_3Ti would be stable and could subsequently nucleate in the α-Al phase, as for alloys whose titanium is found in hyper-peritectic concentrations. Jones [70] supported this titanium gradient theory of segregation, but there is no conclusive evidence that a titanium gradient exists. As early as 1977, Naess and Berg [71] tried to suggest that there would have been a high concentration of titanium around the borides in the liquid pool, but their evidence showed nothing more than the predicted solute profile on the solidification of an Al-Ti alloy.

The duplex theory of nucleation is not totally new. In 1971, Bäckerud [72] claimed to have observed Al_3Ti on boride particles and to have proposed a series of reactions to explain this. It was also mentioned by Cornish [52], who proposed that the role of borides is to facilitate the formation of Al_3Ti at hypo-peritectic concentrations due to a variation in the peritectic composition, which then induces nucleation via the peritectic reaction. They used some of the same arguments as Mohanty et al. [69] about the segregation of titanium to borides. If the TiB_2 particles nucleate the Al_3Ti particles, which, in turn, nucleate the α-Al, then the mechanism is still unexplained, especially since the difference in the expected nucleation temperatures and the non-application of the theory to the alloys of foundries are always a problem. If Al_3Ti particles form on the surface of TiB_2 particles that increase nucleation, then it is the borides that act directly or indirectly as nucleation sites.

1.7 Effect of Si

The alloy composition effect seems to be quite complex. It has been documented for the Al-Si [73–78] system that the grain size first decreases with increasing the alloying concentration and then, after reaching a minimum, the grain size increases with further additions. The minimum is obtained near the maximum solubility limit and some researchers have therefore reported the minimum in grain size at a maximum range of solidification, and therefore at the maximum time of solidification, suggesting that an alloy with a wide range of solidification grants a longer time for nucleation. However, Bäckerud and Johnsson [79] recently suggested that a cellular structure transition to dendrites with well developed orthogonal branches is responsible for the transition, i.e. a cellular-dendritic transition during the growth of equiaxed crystals. They proposed that the transition occurs at a growth restriction factor of 20. This factor is equal to $\Sigma i m_i C_{0,i}$ (k_i-1) where m is the slope of the liquidus line, C_0 is the composition of the liquid and k is the equilibrium distribution coefficient for all elements.

In order to identify the effect of increased silicon content on the morphology and grain size in hypoeutectic Al-Si alloys, Lee et al. [80] used six silicon concentrations, 1, 2, 3, 4, 5 and 8 (wt %) combined with five levels of master alloy (grain refiner) of Al-5%Ti-1%B. The different samples were solidified in a preheated cylindrical graphite crucible at a cooling rate of 0.7°C/s. The results indicate that grain size is controlled by a combination of nucleating power and constitutional conditions at the growing crystal interface. **Figure 20(a)** and **(b)** illustrate the evolution of the grain size according to the concentrations of silicon Si and titanium Ti.

The $TiSi_2$ type phases are the main cause of such poisoning. The micrograph given in **Figure 21** proves the presence of these phases. Other researchers have proposed that they are Al_3Ti particles instead of TiB_2, which become coated by certain complex aluminides or silicides formed by the interaction of silicon with Al_3Ti. These intermetallic compounds may be unable to nucleate aluminum during solidification, since they have always been observed at grain boundaries instead of grain centers [81]. However, since boron atoms are too light to be detected by energy dispersive X-ray spectroscopy, the possibility of the existence of borides cannot be ruled out by the absence of boron in the spectrum. Through X-ray energy dispersive spectroscopy, the $TiSi_2$ phase was identified by the presence of the high intensity peaks relative to titanium and silicon elements. As for **Figure 22**, it illustrates the distribution of the elements of this phase in a 390 alloy treated with 0.4% Ti and cast after 120 minutes. The spectrum relating to the $TiSi_2$ phase is given in **Figure 23**.

1.8 Sr-grain refiner interaction

The addition of strontium as a modifier in Al-Si alloys causes a transition in the morphology of the eutectic silicon from an acicular to a fibrous and fine form. However, strontium can also lead to the formation of a long and columnar α-Al dendritic phase. The Al-5Ti-1B master alloy is often used as a grain refiner to achieve a fine, equiaxed grain structure in aluminum and its alloys. The TiB_2 or/and $TiAl_3$ particles of the Al-5Ti-1B master alloy are thought to be able to act as nuclei for the primary α-Al phase. However, in Al-Si alloys with a high level of silicon (%Si more than 7% by weight), the grain refining power of the master alloy of Al-5Ti-1B is lower compared to that of Al-3B and Al-3Ti-3B master alloys [81, 82]. It was therefore thought that the silicon poisoned the nucleation nuclei. This is mainly related to the formation of silicon and titanium above the $TiAl_3$ particles, and Kori et al. [83] specified that this poisoning effect could be neutralized by increasing the level of addition of the main alloy of Al-5Ti-1B. The sequence of the grain refiner and modifier addition has a significant influence on the grain grade of the α-Al dendritic phase when compared to the combined addition as shown in **Figure 24**.

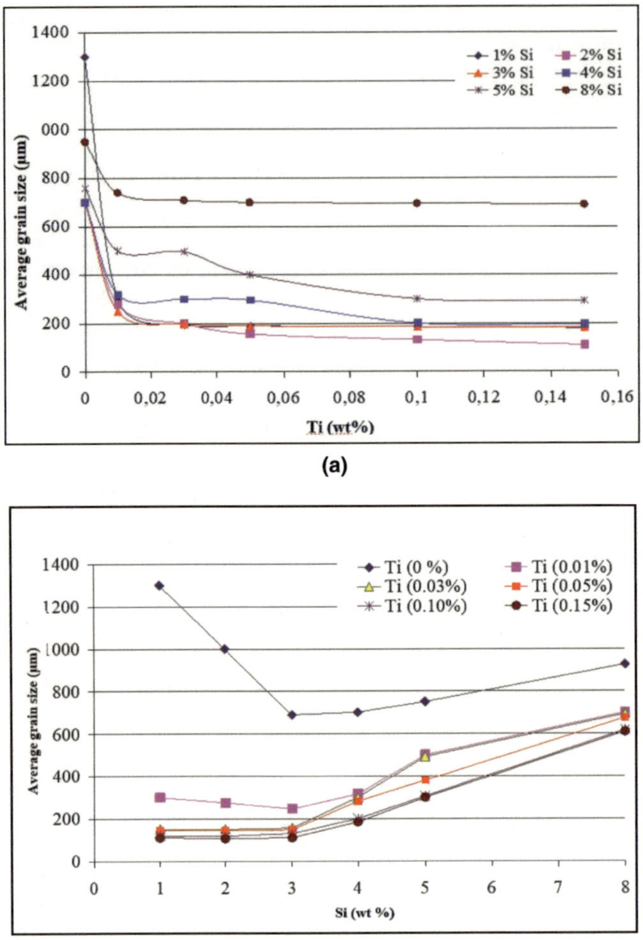

Figure 20.
(a) Average grain size as a function of Ti content for 6 levels of Si, (b) same as in (a) using Al-5%Ti-1%B master alloy.

When the liquid bath is treated with both a grain refiner and a modifier, the evolution of the Sr concentration in the liquid metal is highly time dependent for higher levels of addition of the Al-Ti-B grain refiners, Ti-B **Figure 25** shows this phenomenon. The zero-time concentration is the level of Sr in the liquid aluminum before the addition of either Al-Ti-B grain refiner. After addition, a weakening of Sr is observed for the two liquid alloys; the liquid bath treated with Al-1.5Ti-1.5B master alloy loses its Sr much faster, especially at the initial stage after addition, compared to the liquid bath treated with Al-5Ti-1B. This explains the rapid loss of eutectic modification in the liquid alloy treated with Al-1.5Ti-1.5B, it means that there is insufficient free Sr in the liquid aluminum to modify all the eutectic silicon. The rapid loss of strontium can be explained by external oxidation and vaporization as shown in **Figure 25** [84].

As discussed before, boron B is the most efficient refiner for the A356 alloy. With the addition of strontium and the latter's reaction with boron B, the overall percentages of strontium and boron capable of acting as a refiner and modifier respectively decrease and therefore would not be as effective as those obtained when each is added

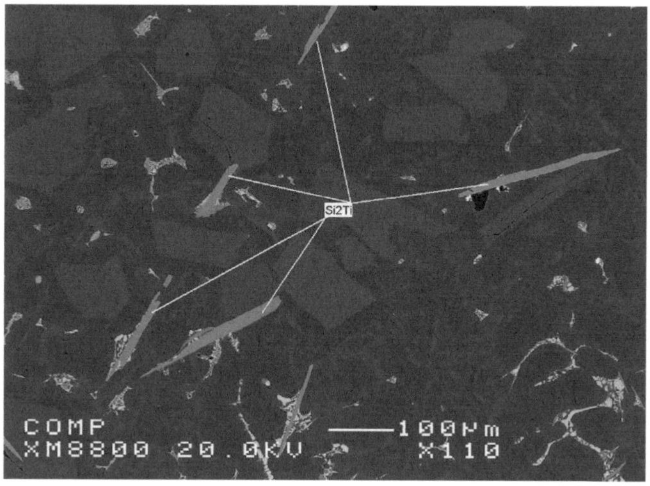

Figure 21.
Backscattered electron micrograph identifying the formation of $TiSi_2$ phases in a 390 alloy (~17%Si) treated with 200 ppm Sr, 0.4%Ti and cast at 750°C after 120 minutes of dwell time.

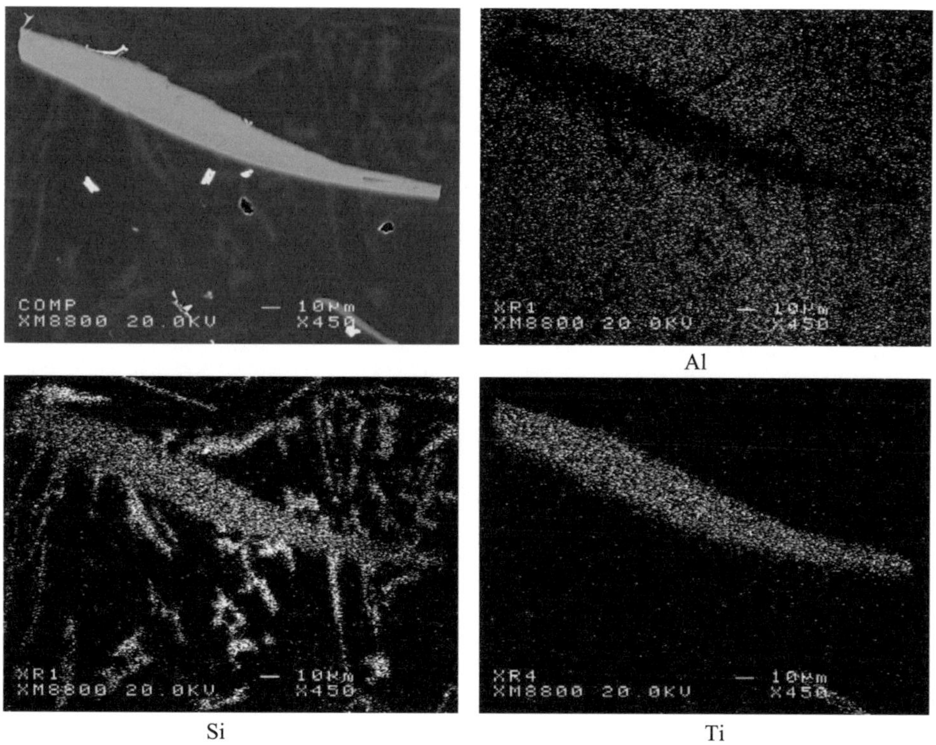

Figure 22.
Distribution of aluminum, titanium and silicon in a 390 alloy treated with Al-10%Ti, 200 ppm Sr, showing the $TiSi_2$ phase and cast after 120 minutes of holding time.

individually. In order to demonstrate the resulting reaction, strontium and boron were added to pure aluminum, and the casting was carried out under the same conditions as those existing during individual addition. Microstructural analysis proved that

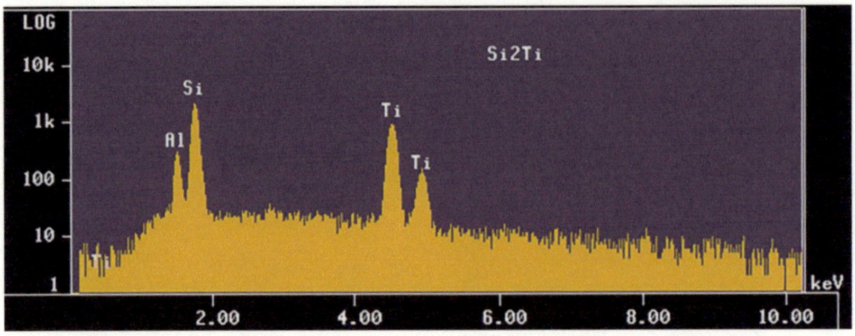

Figure 23.
Dispersive X-ray (EDS) analysis of the $TiSi_2$ phase proven by high intensity peaks.

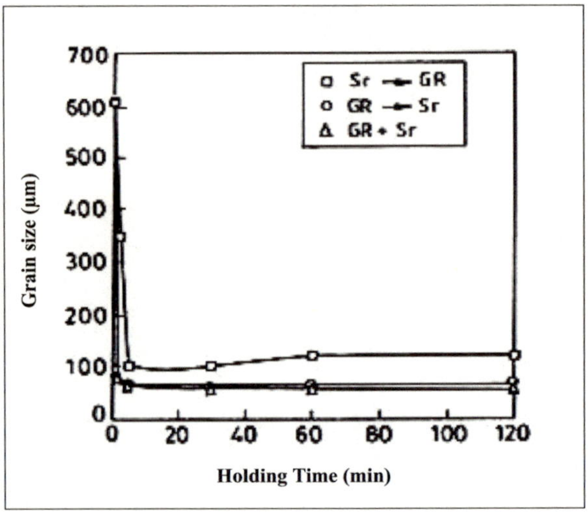

Figure 24.
Analysis of grain size of 356 alloy treated with Al-1%Ti-3%B master alloy: 1%Ti and 0.02%Sr [83].

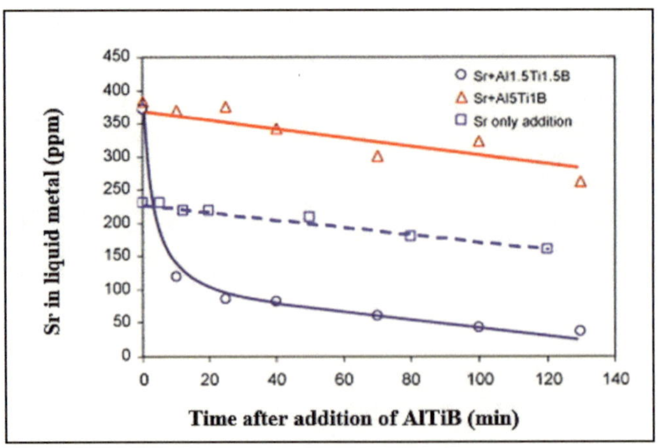

Figure 25.
Sr concentrations as a function of time after AlTiB grain refiners were added to achieve 0.15% Ti in the liquid bath. The broken line refers to the case of Sr modification without addition of any grain refiner.

strontium and boron react with each other forming compounds of the SrB_6 type according to the following reaction: Sr + 6B → SrB_6. This last product is confirmed by the results obtained by Li et al. [85] and Nafisi [86]. This type of compound, whose name is strontium hexaboride, is characterized by a very high melting temperature of 2500°C [87] with a weight ratio of Sr:B equivalent to 1.35:1.

This reaction proves that each atom of Sr could react with six atoms of B and consequently form a compound SrB_6. The intermetallics SrB_6 and TiB_2 could act as nucleants but it should be considered that the consumption of boron in the compound is one of the main parameters. It means that AlB_2 consumes less amount of boron in comparison with SrB_6. Considering a constant amount of boron, the density of nucleating particles is much higher in the case of AlB_2 since a lower number of boron atoms is associated with this compound. Therefore, the greater the number of effective nucleants, the greater the probability of having a smaller grain size. Using dispersive X-ray (EDS) analysis, the SrB_6 phase was identified and confirmed by high intensity peaks. The strong affinity between strontium and boron is shown in **Figure 26** obtained using the electron microprobe. The size of SrB_6 compounds varies between 5 and 10 μm, and their color is a mixture of dark gray and white [88].

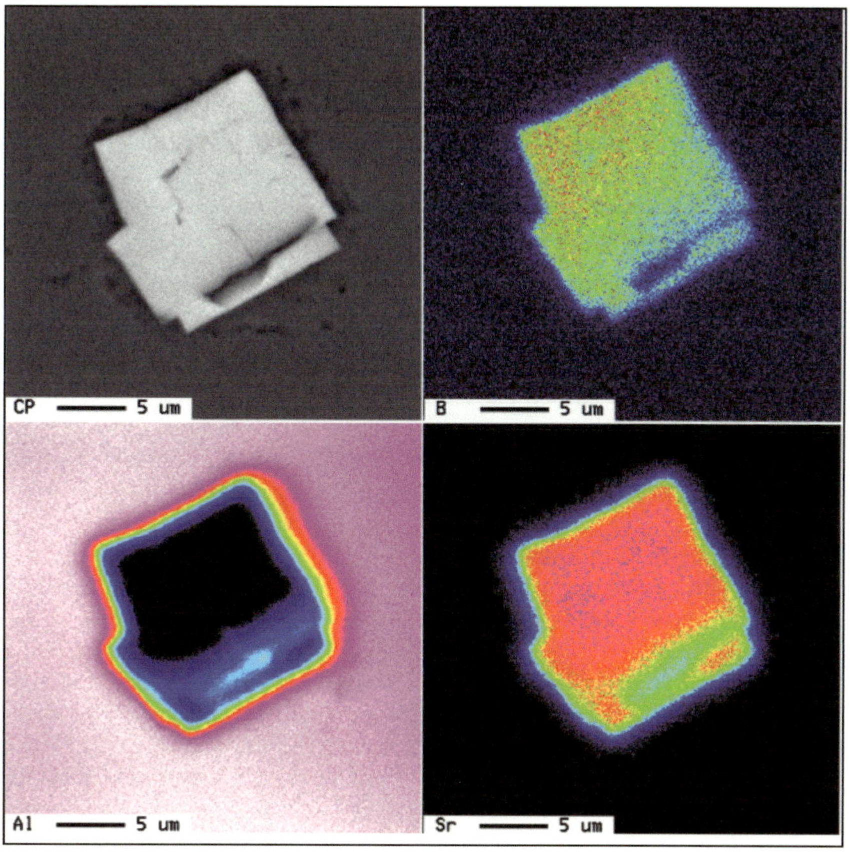

Figure 26.
Mapping produced by the electron microprobe showing the association of strontium and boron in pure aluminum forming SrB_6 type phases [88].

2. Conclusions

Based on the presented survey of literature on the fundamental aspects of grain refining of Al-Si cast alloys, the main points could be summarized as follows:

1. Of primary importance are the grain size, the dimension of the dendritic cells or the interdendritic space, and the shape and distribution of the eutectic mixture, which consists mainly of silicon. A fine, equiaxed grain structure is always desired since a finer grain size promotes improved casting strength by minimizing shrinkage, hot cracking and hydrogen porosity.

2. To achieve grain refining, the most widely practiced way is to introduce effective seed nuclei into the liquid metal using Al-Ti-B grain refiners which usually contain active seeds like $TiAl_3$, TiB_2, AlB_2 or $(Al,Ti)B_2$.

3. The introduction of titanium in Al-10%Ti leads to the formation of ultrafine intermetallics of the $(Al,Si)_3Ti$ type. The latter constitute nucleation sites for the α-Al phase. When added to liquid metal, the Al-4%B master alloy shows remarkable potency in comparison to other grain refiners.

4. For master alloys, residual titanium in alloy A356 reacts with boron B to form TiB_2 which subsequently acts as an active seed alongside AlB_2 for the α-Al phase.

5. The addition of strontium and the grain refiner Al-5%Ti-1%B shows a certain affinity between the modifier and boron. This affinity, limited by the outer surface of the TiB_2, partially deactivates the effect of the refiner since the minimum granular size is obtained for a Ti content of 0.2–0.3% by weight, compared to that obtained with the addition of Al-10%Ti and strontium.

6. The introduction of AlB_2 in Al-4%B form in alloys containing traces of titanium leads to the reaction between boron and titanium to form TiB_2. Grain refining is achieved primarily with TiB_2 rather than AlB_2, or both, depending on the titanium content in the given alloy.

7. In the presence of strontium, boron reacts with strontium to form compounds of the SrB_6 type which is supposed to be a very weak refiner.

8. The affinity between titanium and boron is higher than the affinity existing between boron and strontium. Note also that B does not react with Si unlike titanium, however, it leads to better results.

9. The presence of excess silicon in Al-Si alloys leads to a strong interaction between titanium and silicon. This high affinity leads to the formation of $(Al,Si)_2Ti$-type phases, weakening the nucleation opportunities of the dendritic phase and consequently reducing the degree of grain refinement. Titanium disilicide phase tends to form more when the liquid metal is held for long periods.

Author details

Ehab Samuel[1], Hicham Tahiri[1], Agnes M. Samuel[1], Victor Songmene[2] and Fawzy H. Samuel[1]*

1 Université du Québec à Chicoutimi, Chicoutimi, QC, Canada

2 Department of Mechanical Engineering, École de Technologie Supérieure (ÉTS), QC, Canada

*Address all correspondence to: fhsamuel@uqac.ca

References

[1] Maxwell I, Hellawell A. The constitution of the system Al-Ti-B with reference to aluminum-base alloys. Metallurgical Transactions. 1972;**3**: 1487-1493

[2] Backerud L, Yidong S. Grain refining mechanisms in aluminium as a result of additions of titanium and boron, Part I. Aluminium. 1991;**67**:780-785

[3] Vatne HE. Efficient grain refinement of ingots of commercial wrought aluminium alloys part I: Method for grain refining. Aluminium. 1999;**75**:84-90

[4] Easton M, StJohn D. Grain refinement of aluminum alloys: Part 1. The nucleant and solute paradigms—A review of the literature. Metallurgical and Materials Transactions A. 1999;**30A**: 1613-1633

[5] Jones GP, Pearson J. Factors affecting the grain-refinement of aluminum using titanium and boron additives. Metallurgical Transactions B. 1976;**7B**:223-234

[6] Sigworth GK, Guzowski MM. Grain refining of hypoeutectic Al-Si alloys. AFS Transactions. 1985;**172**:907-912

[7] Kearns MA, Thistlethwaite SR, Cooper PS. Recent advances in understanding the mechanism of aluminium grain refinement by TiBAl master alloys. In: Hale W, edtiors. Light Metals 1996. Warrendale, PA: TMS; 1996. pp. 713-720

[8] Spittle JA, Keeble JM, Al Meshhedani M. A study of grain refinement efficiency in Al-Si alloy castings. In: Proceedings of the 4th Decennial International Conference on Solidification Processing. Sheffield; 1997. pp. 273-276

[9] Dahle AK, Tondel PA, Paradies CJ, Arnberg L. Effect of grain refinement on the fluidity of two commercial Al-Si foundry alloys. Metallurgical and Materials Transactions A. 1996;**27A**: 2305-2313

[10] Fuoco R, Correa ER, de Andrade Bastos M. Effects of grain refinement on feeding mechanisms in A356 aluminum alloy. AFS Transactions. 1998;**106**:401-409

[11] Dahle AK, John DHS, Attavanich P, Taopetch P. Grain formation in AlSi7Mg0.35 foundry alloy at low superheat. Materials Science Forum. 2000;**331–337**:271-276

[12] Bäckerud L, Chai G, Tamminen J. Solidification Characteristics of Aluminium Alloys. In: Foundry Alloys. Vol. 2. Des Plaines, IL, USA: AFS/Skanaluminium; 1990. pp. 71-84

[13] Arnberg L, Bäckerud L, Klang H. Intermetallic particles in Al-Ti-B-type master alloys for grain refinement of aluminium: Part II. Metals Technology. 1982;**9**:7-13

[14] McCartney DG. Grain refining of aluminium and its alloys using inoculants. International Materials Reviews. 1989;**34**(5):247-260

[15] Pearson J, Birch MEJ, Hadlet D. Recent advances in aluminium grain refinement. In: Proceedings of the Conference Solidification Technology in the Foundry and Cast House. London: Metals Society; 1980. pp. 1-5

[16] Volmer M, Weber A. Nucleation in super-saturated products. Zeitschrift für Physikalische Chemie Leipzig. 1925;**119**: 277-301

[17] Yue NL, Lu L, Lai MO. Application of thermodynamic calculation in the in-situ process of Al/ TiB_2. Composite Structures. 1999;**47**: 691-694

[18] Li H, Sritharan T, Lam YM, Leng NY. Effects of processing parameters on the performance of Al grain refinement master alloys Al-Ti and Al-B in small ingots. Journal of Materials Processing Technology. 1997; **66**:253-257

[19] Tuttle BL. Principles of thermal analysis for molten metal process control. In: AFS/CMI Proceedings of the Conference on Thermal Analysis of Molten Aluminum. Rosemont, IL; 1984. pp. 1-36

[20] Mondolfo LF, Barlock JG. Effect of superheating on structure of some aluminum alloys. Metallurgical Transactions B. 1975;**6 B**:565-572

[21] Taylor JA, Wang H, StJohn DH, Bainbridge IF. Anomalous grain coarsening behaviour in grain-refined aluminium alloys cast using low superheat. In: Angier J, editors. Light Metals 2001. Warrendale, PA: TMS; 2001. pp. 935-941

[22] Cook R, Cooper PS, Kearns MA. Benefits of master alloy melt treatments in the aluminum foundry industry. TMS Light Metals. 1996;**1996**:647-654

[23] Cooper P, Hardman A, Boot D, Burhop E. Characterisation of a new generation of grain refiners for the foundry industry. In: Crepeau PN, editor. Light Metals. San Diego, CA: TMS; 2003. pp. 923-928

[24] Lu HT, Wang LC, Kung SK. Grain refining in A356 alloys. Journal of Chinese Foundryman's Association. 1981;**29**:10-18

[25] Hayes FH, Lukas HL, Effenberg G, Petzow G. Thermodynamic calculation of the Al-Rich corner of the Al-Ti-B System. Zeitschrift für Metallkunde. 1989;**80**:361-365

[26] Roger P, Bauer J, Bohn M. Internal report, University of Vienna. 1996

[27] Zupanic F, Spaic S, Krizman A. Contribution to ternary system Al-Ti-B part 2 - study of alloys in Al-AlB_2-TiB_2 triangle. Materials Science and Technology. 1998;**14**:1203-1212

[28] Zupanic F, Spaic S, Krizman A. Contribution to ternary system Al-Ti-B Part 1: Study of diborides present in the aluminium corner. Materials Science and Technology. 1998;**14**:601-607

[29] Abdel-Hamid A, Hamar-Thibault S, Durand F. Nature and morphology of crystals rich in Ti and B in Al-Rich Al-Ti-B alloys. Journal of Crystal Growth. 1984;**66**:195-204

[30] Fjellstedt J, Jarfors AEW, Svendsen L. Experimental analysis of the intermediary phases AlB~2 AlB~1~2 and TiB~2 in the Al-B and Al-Ti-B systems. Journal of Alloys and Compounds. 1999; **283**:192-197

[31] Gröbner J, Mirkovic D, Schmid-Fetzer R. Thermodynamic aspects of grain refinement of Al-Si alloys using Ti and B. Materials Science and Engineering A. 1995;**395**:10-21

[32] Youdelis WV, Yang CS. Ti(Al,Si)3 compound formation in non-equilibrated Al-Ti-Si. Metal Science. 1980;**14**:500-501

[33] Sokolowski JH, Kierkus CA, Brosnan B, Evans WJ. Formation of insoluble Ti (Al,Si)3 crystals in 356 Alloy casting and their sedimentation in foundry

equipment: Causes, effect and solutions. AFS Transactions. 2000;**108**:491-496

[34] Ramos AS, Nunes CA, Rodrigues G, Suzuki PA, Coelho GC, Grytsiv A, et al. Ti6Si2B a new ternary phase in the Ti-Si-B System. Intermetallics. 2004;**12**: 487-491

[35] Seifert H-J, Lukas HL, Petzow G. Thermodynamic optimization of the Ti-Si system. Zeitschrift für Metallkunde. 1996;**87**:2-13

[36] Youdelis WV, Yang CS. Non-peritectic grain refinement of aluminum by titanium. Aluminium. 1980;**1980**: 411-413

[37] Sigworth GK. Theoretical and practical aspects of the modification of Al-Si alloys. AFS Transactions. 1983;**91**: 7-16

[38] Simensen CJ. Grain refining of Al-7wt%Si alloys. The Minerals, Metals & Materials Society. 1999;**1999**:679-684

[39] Li P, Kandalova EG, Nikitin VI. Grain refining performance of Al-Ti master alloys with different microstructures. Materials Letters. 2005; **59**:723-727

[40] Tøndel PA, Arnberg L. Grain refinement of an Al-10%Si alloy by Ti-additions. In: The 3rd International Conference on Aluminium Alloys. pp. 129-134

[41] Quested TE. Understanding mechanisms of grain refinement of aluminium alloys by inoculation. Materials Science and Technology. 2004; **20**(11):1357-1369. DOI: 10.1179/026708304225022359

[42] Wang X, Liu Z, Dai W, et al. On the understanding of aluminum grain refinement by Al-Ti-B type master alloys. Metallurgical and Materials Transactions B. 2015;**46**:1620-1625. DOI: 10.1007/s11663-014-0252-3

[43] Mirkovic D, Gröbner J, Schmid-Fetzer R, Fabrichnaya O, Lukas HL. Experimental study and thermodynamic re-assessment of the Al-B System. Journal of Alloys and Compounds. 2004; **384**:168-174

[44] Kori SA, Murty BS, Chakraborty M. Development of an efficient grain refiner for Al-7Si alloy and its modification with strontium. Materials Science and Engineering A. 2000;**283**:94-104

[45] Wang L, Bian X. Refining effect of boron on hypoeutectic Al-Si alloys. Journal of Materials Science and Technology. 2000;**16**(5):517-520

[46] Kori SA, Murty BS, Chakraborty M. Development of an efficient grain refiner for Al-7Si Alloy. Materials Science and Engineering A. 2000;**280**:58-61

[47] Easton MA, Qian M, Prasad A, StJohn DH. Recent advances in grain refinement of light metals and alloys. Current Opinion in Solid State and Materials Science. Feb 2016;**20**(1): 13-24

[48] Mondolfo LF, Farooq S, Tse C. Grain refinement of aluminium alloys by titanium and boron. In: Solidification Processing 1987. Sheffield: The Institute of Metals; 1988. pp. 133-136

[49] Crossley FA, Mondolfo LF. Mechanism of grain refinement in aluminium alloys. Journal of Metals – Transactions of the American Institute of Mining and Metallurgical Engineers. 1951;**191**:1143-1148

[50] Hu B, Li H. Grain refinement of DIN226S alloy at lower titanium and boron addition levels. Journal of

Materials Processing Technology. 1998; **74**(1):56-60

[51] Li H, Chandel RS, Sritharan T. Effect of Al-Ti and Al-B master alloy addition on the grain refinement of stationary arc-melted Al weld. Journal of Materials Science Letters. 1996;**15**:1886-1887

[52] Cornish AJ. The influence of boron on the mechanism of grain refinement in dilute aluminium-titanium alloys. Metal Science. 1975;**9**:477-484

[53] Cibula A. The mechanism of grain refinement of sand casting in aluminium alloys. The Journal of the Institute of Metals. 1949–1950;**76**:321-360

[54] Kiusalaas R, Bäckerud L. Influence of production parameters on performance of Al-Ti-B master alloys. Solidification Processing, The Institute of Metals. 1988;**1988**:137-140

[55] Schumacher P, Greer AL. Studies of the action of grain-refining particles in aluminium alloys. In: Grandfield JF, Eskin DG, editors. Essential Readings in Light Metals. Cham: Springer; 2016. DOI: 10.1007/978-3-319-48228-6_44

[56] Davis IG, Dennis JM, Hellawell A. Metallurgical Transactions. 1970;**1**: 275-279

[57] Maxwell I, Hellawell A. A simple model for grain refinement during Solidification. Acta Metallurgica. 1975; **23**:229-237

[58] Arenberg L, Bäckerud L, Klang H. Metall Technologie. 1982;**2**:465

[59] Vader M, Noordegraaf J. Light Metals. In: Campell PG, editor. Warrendale, PA: TMS; 1989. pp. 937-941

[60] Bäckerud L, Gustafson P, Johnsson M. Grain refining mechanisms in aluminum as a result of additions of titanium and boron, part II. Aluminium. 1991;**67**(9):910-915

[61] Bäckerud L, Krol E, Tamminen J. Solidification characteristics of aluminium alloys. In: Wrought Alloys. Vol. 1. Oslo, Norway: AFS/ Skanaluminium; 1986. pp. 34

[62] Johnsson M, Bäckerud L, Sigworth GK. Study of the mechanism of grain refinement of aluminium after additions of Ti- and B-containing master alloys. Metallurgical Transactions A. 1993;**24A**: 481-491

[63] Johnsson M. On the mechanism of grain refinement of aluminium after additions of Ti and B. In: Light Metals 1993. Warrendale, PA: The Minerals, Metals and Materials Society; 1993. pp. 769-777

[64] Samuel FH. Studies on addition of inclusions to molten aluminum using a novel technique. Metallurgical and Materials Transactions B. 1995;**26B**: 103-109

[65] Easton MA, John DHS. A model of grain refinement incorporating alloy constitution and potency of heterogeneous nucleant particles. Acta Materialia. 2001;**49**:1867-1878

[66] Guthrie RIL. Studies on the fading behavior of Al-Ti-B master alloys and grain refinement mechanism using LiMCA. Light Metals. 1995;**1995**:859-868

[67] Mohanty PS, Samuel FH, Gruzleski JE. Studies on addition of inclusions to molten aluminum using a novel technique. Metallurgical and Materials Transactions B: Process Metallurgy and Materials Processing Science. 1995;**26**(1):103-109

[68] Mohanty PS, Samuel FH, Gruzleski JE, Kosto TJ. Studies on the mechanism

of grain-refinement in aluminium. In: Mannweiler U, editor. Light Metals 1994: Proceedings of the Technical Sessions Presented by the TMS Light Metals Committee at the 123rd TMS Annual Meeting, San Francisco, February 27–March 3, 1994. San Fransisco, California, USA; 1994. pp. 1039-1045

[69] Mohanty PS, Samuel FH, Gruzleski JE. Mechanism of heterogeneous nucleation of pores in metals and alloys. Metallurgical Transactions A: Physical Metallurgy and Materials Science. 1993; **24**(8):1845-1856

[70] Jones GP. Grain refinement of castings using inoculants for nucleation above liquidus. Solidification Processing, The Institute of Metals. 1987;**1987**:496-499

[71] Naess SE, Berg O. Zeitschrift für Metallkunde. 1974;**65**:599-602

[72] Bäckerud L. On the grain refining mechanism in Al-Ti-B alloys. Jernkontorets Ann. 1971;**155**:422-424

[73] Samuel AM, Mohamed SS, Doty HW, Valtierra S, Samuel FH. Some aspects of grain refining of Al-Si cast alloys. International Journal of Cast Metals Research. 2019;**32**(1):1-14

[74] Samuel AM, Mohamed SS, Doty HW, Valtierra S, Samuel FH. Grain refining of Al-Si alloys using Al-10% Ti master alloy: Role of Zr addition. International Journal of Cast Metals Research. 2019;**32**(1):46-58

[75] Samuel AM, Doty HW, Valtierra S, Samuel FH. A metallographic study of grain refining of Sr-modified 356 alloy. International Journal of Metalcasting. 2017;**11**(2):305-320

[76] Tahiri H, Mohamed SS, Doty HW, Valtierra S, Samuel FH. Effects of grain refining on columnar-to-equiaxed transition in aluminum alloys. In: Sivasankaran S, editor. Aluminium Alloys - Recent Trends in Processing, Characterization, Mechanical Behavior and Applications. London, UK: Intech; 2017

[77] Samuel AM, Doty HW, Valtierra S, Samuel FH. Effect of grain refining and Sr-modification interactions on the impact toughness of Al–Si–Mg cast alloys. Materials & Design. 2014;**56**: 264-273

[78] Habibi N, Samuel AM, Samuel FH, Rochette P, Paquin D. Effect of grain refining and Sr modification on Prefil measurement sensitivity in 356 alloys using electron probe microanalysis technique. International Journal of Cast Metals Research. 2004;**17**(2):79-87

[79] Bäckerud L, Johnsson M. The relative importance of nucleation and growth mechanism to control grain size in various aluminum alloys. In: Hale W, editor. Light Metals 1996. Warrendale, PA; TMS; 1996. pp. 679-685

[80] Lee YC, Dahle AK, StJohn DH, Hutt JEC. The effect of grain refinement and silicon content on grain formation in hypoeutectic Al-Si alloys. Materials Science and Engineering A. 1999;**259**: 43-52

[81] Rao AA, Murty BS, Chakraborty M. Influence of chromium and impurities on the grain-refining behavior of aluminum. Metallurgical and Materials Transactions A. 1996;**27**:791-800

[82] Sritharan T, Li H. Influence of titanium to boron ratio on the ability to grain refine aluminium-silicon alloys. Journal of Materials Processing Technology. 1977;**63**:585-589

[83] Kori SA, Murty BS, Chakraborty M. Development of an efficient grain refiner

for Al–7Si alloy and its modification with strontium. Materials Science and Engineering. 2000;**83**(1–2):94-104

[84] Lu L, Dahle AK. Effects of combined additions of Sr and AlTiB grain refiners in hypoeutectic Al-Si foundry alloys. Materials Science and Engineering A. 2006;**435–436**:288-296

[85] Li JG, Zhang BQ, Wang L, Yang WY, Ma HT. Combined effect and its mechanism of Al-3wt.%Ti-4wt.%B and Al-10wt.%Sr master alloy on microstructures of Al-Si-Cu alloy. Materials Science and Engineering A. 2002;**A328**:169-176

[86] Nafisi S. Effects of grain refining and modification on the microstructural evolution of semi-solid 356 Alloy. [PhD thesis]. Chicoutimi, Canada: Université du Québec à Chicoutimi (UQAC); 2006. p. 358

[87] Massalski TB, Scott WW. Binary Alloy Phase Diagrams. 2nd ed. USA; 1990. p. 540

[88] Paradis M, Abdelaziz MH, Doty HW, Samuel FH. On the mechanical properties of lost foam Cast A356 automotive components: Effects of melt treatment and solidification conditions. International Journal of Metalcasting. 2017;**11**(3):494-505

Section 2

Fabrication of Aluminum Alloys

Chapter 3

Quality Assurance of Aluminium Extrusion for 6xxx Series Alloys

Ying Pio Lim and Heng Kam Lim

Abstract

Aluminium extrusion of 6xxx series alloys is gaining more and more importance and indispensable in the market for applications in automotive (great potential for EV in the near future by 2030), construction, architecture, electronics, marine and rail transport. The 6000 series alloys can be divided into soft alloy (e.g. 6060, 6063) and hard alloy (e.g. 6005A, 6061, 6082) for different applications based on customer's requirements for tensile strength, yield strength, elongation, surface finishing (powder coating and anodizing) and heat treatment. To produce good quality extrudates with quality that can meet customer's stringent requirements has become a challenging job nowadays for extruders in developing country like Malaysia. In order to be competitive in the global market, the products have to be produced at minimum cost and just-in-time to meet the committed delivery date. This will require a very good implementation of quality system in the production to ensure customer's satisfaction is achieved from time to time. Based on the real experiences of working in an international scale extruder, the effective methods taken to improve product quality and productivity are elaborated throughout the chapter.

Keywords: aluminium alloy, extrusion, defects, quality assurance, material handling

1. Introduction

Aluminium alloy can be defined as a substance having metallic properties and composed of two or more alloying elements of which the base metal is aluminium. Most aluminium alloys contain 90–96% aluminium, with one or more other elements added to provide a specific combination of properties and characteristics. It is quite usual to have several minor alloying elements in addition to one or two major alloying elements to impart special fabrication or performance characteristics for the sake of manufacturability and desired mechanical properties. The 6xxx series alloys have both magnesium and silicon as their main alloying elements, which combine as magnesium silicide (Mg_2Si) following solid solution [1]. Alloys in this series are heat treatable. This series of aluminium alloys can be divided into soft alloy and hard alloy (jargon used by production people to indicate the tensile strength of the alloys, soft alloy has maximum tensile strength 215 MPa, while hard alloy has maximum tensile strength 310 MPa). Examples of soft alloy are 6063, 6061 and 6463, while the hard alloy encompasses 6061, 6005A and 6082 for the most common applications in consumer products, architectural structures and construction. **Figure 1** indicates

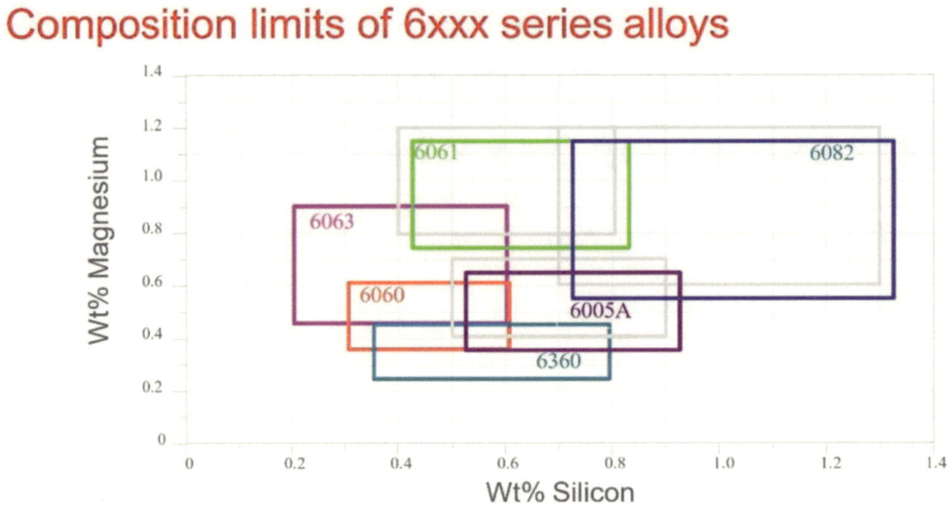

Figure 1.
6xxx series alloy types and their Magnesium and Silicon weight percentages.

the category of 6xxx series alloys according to their silicon and magnesium contents, which corresponds to their mechanical properties.

The 6xxx series alloys are especially suitable for hot extrusion process because of its extrudability to form solid or hollow/semi-hollow cross sections. The alloys are also heat treatable with artificial aging process to achieve the desired tensile strength, yield strength and elongation. There are a number of reasons why the 6xxx series are popular for various applications in extruded profiles, as stated below [2]:

- Density approximately one-third that of steel and has high strength-to-weight ratio with tensile strength ~340 MPa for 6082-T6 alloy.
- Good electrical conductivity of 48–50 IACS for alloy containing ≥97.6% Al and ≤0.19% Fe.
- Good thermal conductivity to be made as heat sink.
- Good corrosion resistance.
- Good ductility and workability (due to the fcc structure) for fabrication by rolling, stamping, drawing, spinning, roll forming, forging, and extrusion.
- Cryogenic toughness, as the fcc structure does not become brittle at low temperatures.
- Variety of surface finishes ranging from clear to colour anodized, colour powder coating and PVDF coating (polyvinylidene fluoride) for functional or cosmetic applications.
- Nontoxic for food storage, cookware, and food processing applications (RoHS and REACH Compliance).
- Recyclable for sustainable manufacturing.

In this book chapter, the focus is on conventional direct hot extrusion process. The word "hot" is referring to the preheating of the billet before the direct extrusion process. The basic process consists of forcing a preheated billet of round shape which is loaded into a container by a hydraulic ram with dummy block at its front end. The ram is linked to the main piston of a main cylinder powered by hydraulic system [3]. The aluminium billet can be preheated from 400 to 500°C in a preheat-oven designed with 3–4 heating zones. The temperature settings are in increasing trend from inlet zone to outlet zone. The billet will be forced to squeeze through a die which is also preheated to about 450–480°C to form a uniform cross-section profile either in solid, hollow or semi-hollow shape according to specific product design. The extruded profile will be subjected to air- or water-cooling process according to its temper requirements [4]. The run-out table on which the profile is placed is installed with cooling fans to blow the profile for continuous cooling process until it is stretched to straighten the material in the long length (ranging from 20 to 40 m). The stretching process also serves to impose strain hardening effect on the material for subsequent effective natural and artificial aging to achieve the desired mechanical properties. The stretched profile will be moved by conveyor to the cutting station to be cut into the desired order length (with tolerance of ±5 mm) and the material will be loaded into trolley and stacked into layers separated by spacer bars.

2. Defects in extrusion

Aluminium alloy in extruded form is easily subject to damage due to external forces because the surface is relatively soft and requires extra precaution to prevent the unwanted damage that can cause the extruded profile deemed to be a reject (unacceptable product that does not fulfil standard or customer specifications). The defects in extrusion can be caused by multiple factors in the process related to the 4Ms in root cause analysis technique, namely man, machine, material and method. There are four main reasons that contribute to the defect's formation in the extruded products [5].

i. Defects in incoming billet (e.g. hydrogen contents exceed 0.2 cc/100 g, lack of homogenization—β to α transformation <80%, inclusion, scales/flakes, internal crack, undissolved oxides etc.).

ii. Physical defects in die-set (e.g. mandrel offset, die bearing worn-out, die aperture oversized cause dimensional inaccuracy etc.).

iii. Extrusion process parameters not optimized (e.g. inappropriate extrusion pressure, billet temperature, ram speed etc.).

iv. Post-extrusion operations suboptimization (e.g. rough material handling, insufficient stretching, saw chips at cutting machine caused scratches etc.).

A local extruder from Malaysia (company's name is not to be disclosed) is willing to share with the authors the extrusion rejection data from Jan till Sept 2022. The data is shown in **Figure 2**.

The top 10 defects in extrusion in descending order are dented surface, bubble/blister, tearing, scratches, die broken, water mark, streaking line, shape out, backend defect and soda mark.

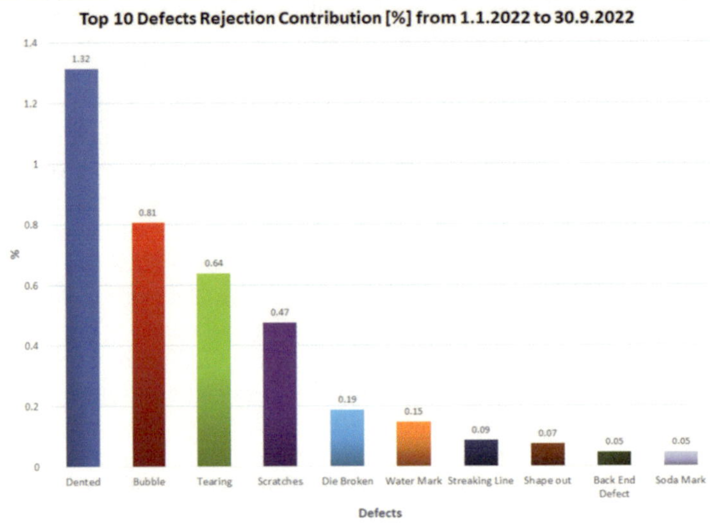

Figure 2.
Top 10 defects of extrusion for 9 months.

To ease understanding, each of the defects are briefly explained below and illustrated with pictures.

i. **Dented**. Depressions appeared generally on the runout face of the extruded profile, **Figure 3**.

ii. **Bubble/blister**. Raised areas on the profile surface most often aligned in the extrusion direction due to subsurface gas expansion, **Figure 4**.

iii. **Tearing**. Fine transverse cracks or tears associated with areas of high friction such as edges. They occur when exit temperature or speed is too high, **Figure 5**.

Figure 3.
Dented mark on extrudate.

Figure 4.
Bubble or blister appearing on extrudate surface.

Figure 5.
Tearing.

iv. **Scratches**. Superficial marking on surface due to interfacial contact between aluminium and other materials, **Figure 6**. Scratches are undesired in milled finished (MF) products if good surface finishing is required. Scratches which is not deep (<0.3 mm) are acceptable for products that will be powder coated because it can be covered. Light and heavy scratches (above 0.1 mm) are rejected for products that need anodizing process because the anodic layer is too thin (25 micron max) to cover the scratches. Surface that is subjected to sand blasting and anodizing might not be able to eliminate the scratches too.

v. **Die broken**. When die is broken at the die plate bearing area, a non-perfect cross-sectional shape will be formed. The defect will be continuous along longitudinal direction of extrusion, **Figure 7**.

Figure 6.
Scratches.

Figure 7.
Die broken.

vi. **Water mark**. Heavy oxidation of surface associated with entrapment of moisture. Could be due to direct contact with rain water, residual cooling water from extrusion machine's cooling box or condensation, **Figure 8**.

vii. **Streaking lines**. They are bands or lines appearing darker or lighter, brighter or duller, in colour and tone different from the remainder of the surface, **Figure 9**.

Figure 8.
Water mark.

Figure 9.
Streaking lines.

The basic cause of this streaking is a difference in microstructure between the streaked portion of the extrudate surface and the remainder, which leads to a difference in response in etching and anodizing.

viii. **Shape out**. Basically, it is dimension out of spec, linear or angular dimension, **Figure 10**. This is more inclined to happen on hollow profiles than solid profiles. Deflection of mandrel due to high pressure is one of the major causes.

Figure 10.
Shape out.

ix. **Backend defect**. A coned shape defect formed at one third of the extrudate due to the centre material of billet flowing through the die while the billet outer skin remains stationary at the container wall. This forms an annular separation in the cross section where there is a separation of inner core and external zone. After anodizing, it forms dull line underneath the surface, **Figure 11**.

x. **Soda mark**. It is not corrosion or water mark but it will appear after anodizing. If material was delayed for rinsing after etching, the residual caustic soda on the surface can cause soda mark, **Figure 12**.

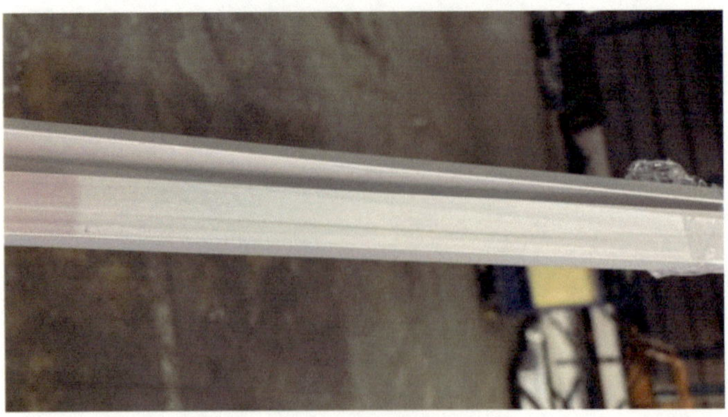

Figure 11.
Backend defect.

Figure 12.
Soda mark.

3. Customer's requirements for extrudates

As an extruder who is highly reputable for its quality products that can meet the requirements of global customers with competitiveness in quality, delivery and after-sales service, the company is used to gaining the customer's confidence and provide quality assurance by signing agreement with customer for the high volume of orders received. The agreement entrusts the company to manufacture and deliver the aluminium profiles all according to the terms and conditions contained in the agreement. The company must be ISO 9001 certified and undertakes to manufacture the products in strict compliance with all provisions of the agreement signed. Therefore, it is necessary for the manufacturer to understand the quality requirements of the customer thoroughly and execute effective quality management system to fulfil the requirements to ensure customer's satisfaction is achieved to secure long term business relationship.

The extruded products can be categorized into three main categories, namely mill finish profiles, powder coated profiles and anodized profiles. In this chapter, the focus is on mill finish profiles. The mill finish profiles to be used in painting and anodizing applications are required to comply with standards EN 12020-2, EN 755 and EN 573. The dimensional, mechanical and surface aspect requirements are usually specified in the product drawings and quality documents provided by customers.

Aluminium profiles are not only meant to work as structural support, but more importantly they are also for decorative purpose. Therefore, cosmetic criteria for surface of the profile are very important. The surface inspection can be categorized into three types:

- Primary/main—Visually critical surface directly viewed by observer and forms part of the product function, e.g. door frame.

- Secondary—Visible but not critical because observer does not view it directly, but it shall be of uniform surface quality, e.g. door edges.

- Non-visible—Non visible surface which is hidden from normal observation angle.

Typical surface defects characteristics are listed in **Table 1** with their inspection criteria.

Every extruded profile has a specific material specified in the drawing. The correct material must be used (e.g. 6063) and its chemical compositions must be verified by spark test (using arc/spark optical emission spectrometry (OES) analyser). The chemical compositions must comply with EN 755-2 or ASTM B221-14. The customer has the right to cut sample from the delivered lots and send to a third-party laboratory to verify its chemical compositions. Dispute will arise if the compositions vary from the results stated on the mill-cert of supplier. In that situation, the sample will be sent back to supplier for inhouse testing and another third-party testing.

No	Defect characteristic	Direct view quality	Indirect view quality
1	Spot defects (crater, spots, solid-, slag-, oxide-inclusion, pick-up, etc.)	Max 5 defects per meter/per side with a diameter of < 1.0 mm. Defects allowed only if not contrasting appearance. Distance of defects must not be located closer than 100 mm to each other in all sides/direction.	Max 7 defects per meter/per side with a diameter of < 1.0 mm. Defects allowed only if not contrasting appearance. Distance of defects must not be located closer than 100 mm to each other in all sides/direction.
2	Scratches (linear, mechanical damage of the surface)	Max 5 defects per meter/per side, wide, with a diameter of <0.15 mm, length < 15 mm. Defects allowed only if not contrasting appearance. Distance of defects must not be located closer than 100 mm to each other in all sides/direction.	Max 7 defects per meter/per side, wide, with a diameter of <0.15 mm, length <15 mm. Defects allowed only if not contrasting appearance. Distance of defects must not be located closer than 100 mm to each other in all sides/direction.
3	Stains and discoloration	Not allowed.	Not allowed.
4	Roughness, cracks	Not allowed.	Not allowed.
5	Extrusion line (groove/grooving)	Allowed only by regular optical appearance.	Allowed only by regular optical appearance.
6	Mechanical damage of all forms (dented, chipped off, etc.)	Not allowed.	Not allowed.
7	Semi finish related irregularities (bumps, lumps, streaking lines, etc.)	Not allowed.	Not allowed.
8	Water corrosion	Not allowed.	Not allowed.
9	Stop marks (band like pattern visible around the full perimeter)	Not allowed.	Not allowed.
10	Tearing/speed cracks (initial crack at edge of die bearing and propagates to extruded part)	Not allowed.	Not allowed.

Table 1.
Extrusion surface defects characteristic.

Mechanical property is an essential requirement of the extruded profiles. The specifications of mechanical property can be found in EN 755-2 or ASTM B221-14. Usually, the supplier has to possess a calibrated static tensile test machine of 100 kN capacity to do the test internally. On special request, the test can be done externally by a certified test service provider and certified test report is produced. The test results will be included in the mill-cert of the product. Usually the ultimate tensile strength, yield strength and elongation will be reported to qualify the product. The test specimen's dimensions must follow international standard such as that specified in ASTM B557M-15. The mechanical property is related to the temper of the alloy. 6xxx series aluminium alloy is heat-treatable. The correct process must be done to achieve the desired temper. When the products are supplied for marine applications, the mechanical property requirements will have to comply with either Bureau Veritas (BV) Rules on Materials and Welding for the Classification of Marine Units NR216 or American Bureau of Shipping (ABS) Rules for Materials and Welding (Part 2). The standards have specific test specimen's dimensions different from that of ASTM. The tests required to do include tensile test and drift expansion test (to test compression strength). The most common materials used for marine applications are 6061-T6 and 6082-T6.

Dimensional tolerances of the extruded profiles are critical to meet the customer requirements as first priority. Malaysia's extruders are used to following the JIS H 4100 standard for dimensional tolerances. The standard provides clear guidance on linear length, angle, straightness, flatness and twist. Advanced measuring machine like Romidot Vision H300 is used to measure the linear and angular dimensions of the profiles. The profiles have to be cut and deburred before measurement. Measurement of straightness, flatness and twist has to be done on a granite measuring table that has to be calibrated for levelling.

Quality can be defined by the degree of consumer satisfaction where the products are produced according to all technical specifications stipulated on drawings and customer-supplier quality agreement. Quality is also considered as faultless products where fewer defects is equivalent to lower costs. Hence, to satisfy customer needs and ensure product delivery according to their requirements, it is necessary to find solutions to overcome quality issues by gathering information about the entire production chain, analysing it, and making better decisions to implement continuous improvement by using PDCA methodology. It can help companies improve their operational efficiency and overall product quality [6].

4. Quality assurance activities in extrusion

The company is specialized in the development and production of aluminium profiles for applications in engineering, architectural and industrial works in general. The products are supplied to local market and also to global market in South East Asia, North America, Europe and Australia. The company's quality policy is aimed at absolute customer satisfaction with punctual delivery and meeting the product functional requirements. The company's pursuit of excellence is a constant pursuit to strengthen business relationship, committed to developing and continuously improving the product quality and satisfactory after sales service. Therefore, Quality Management System (QMS) is its central pillar and various quality assurance activities have to be planned and implemented systematically to achieve the quality objectives. The following sub-sections will elaborate activities that have been implemented in the production system to ensure the right product quality at the right cost.

4.1 Control plan

A control plan describes the methods for controlling product and process variation in order to produce quality parts that meet customer requirements. Control plans are a critical part of the overall quality process. They are living documents that are updated as processes change and improve throughout the product lifecycle [7]. The product control plan consists of process flow in its second column; therefore, it is also considered as process control plan incorporated. The control plan is designed specifically for a customer who has stringent quality requirements and they will conduct supplier audit to confirm that their products have product/process control plan to ensure good quality. Technical specifications are specified in the control plan which include process parameters, QC inspection criteria and chemical compositions. Control points are the location where measurement is done and specific equipment is listed. Related document or record is also specified and responsible persons are stated. **Figure 13** shows an extracted example of control plan (specific data is obscured for confidential purpose).

4.2 FMEA

Failure Mode and Effects Analysis (FMEA) is a guide to the development of a complete set of actions that will reduce risk associated with the system, subsystem, and component or manufacturing or assembly process to an acceptable level [8]. The FMEA concerned here is Process FMEA (PFMEA) which is used to analyse the already developed or existing processes. PFMEA focuses on potential failure modes associated with both the process safety/effectiveness/efficiency, and the functions of a product caused by the process problems. PFMEA is a structured approach designed to achieve the following objectives:

- Predict failures and prevent their occurrence in manufacturing and other functional areas that generate defects.

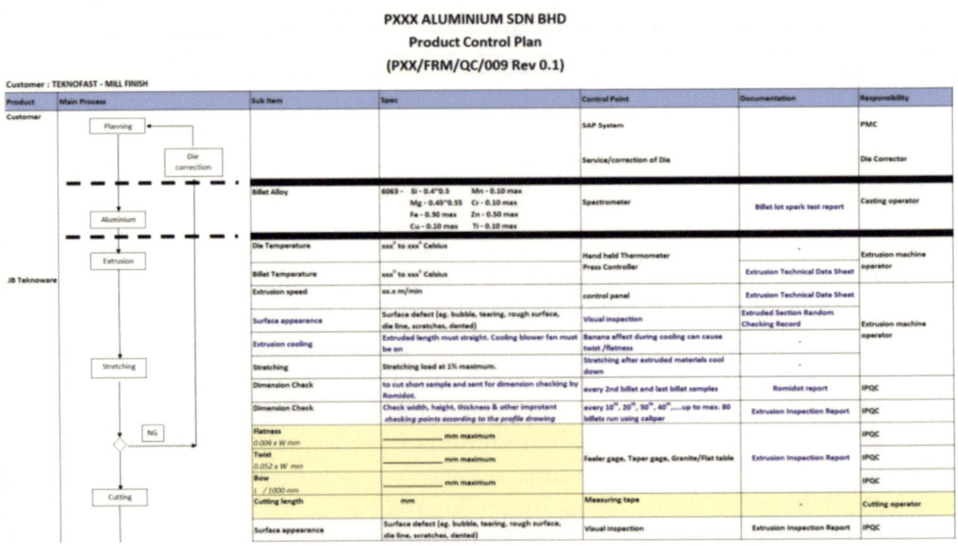

Figure 13.
Product control plan.

- Identify the ways in which a process can fail to meet critical customer requirements.
- Estimate the Severity, Occurrence and Detection (SOD) of defects.
- Evaluate the current Process/Product Control Plan for preventing these failures from occurring and escaping to the customer.
- Prioritize the actions that should be taken to improve and control the process using a Risk Priority Number (RPN).
- Minimize loss of product performance or performance degradation.
- Develop Preventive Maintenance plans for in-service machinery and equipment.

A partially extracted example of PFMEA for extrusion process is shown in **Figure 14**.

Failure Mode and Effect Analysis (FMEA)

Title	Extrusion Process at 7B	**Department**	Extrusion	**Approval**	
Authors	Mhd Noor	**Revision Date**	1/9/2020		
		Revision No.	0		

Process	Failure Mode	Failure Effect	Severity (S)	Criticality (SxO)	Root Cause	Occurrence (O)	Current Control	Detection (D)	RPN (SxOxD)	Actions Recommended	Responsible Person	Evaluation After Actions: Severity (S)	Occurrence (O)	Criticality (SxO)	Detection (D)	RPN (SxOxD)
What is the process step or feature under investigation?	In what ways could the step or feature go wrong?	What is the impact on the customer (internal/external) if this failure happened?			What causes the step or feature to go wrong? (how could it occur?)		What controls exist that either prevent or detect the failure?			What are the recommended actions for reducing the occurrence of the cause or improving detection?	Who is responsible for making sure the actions are completed?					
Die Preparation	Using wrong or unqualified die	Dimensional rejection at Fabrication	7	16	Die number and copy number on dies difficult to see.	2	Visual check by Tooling worker	2	32							
		Failure to assembly at customer site	8													
	Die in oven less than 4 hours.	Die MC during extrusion	6	30	No other dies available for extrusion or too late put dies into oven.	5	Record time of die put into the oven and extrusion time	4	120	Install screen monitor to shows the time in and duration in oven. Alert by colour any less and over duration.	Extrusion HOD & Maintenance (June 2020)	6	2	12	4	48
	Oven temperature setting lower than 460°C.	Die MC during extrusion	6	6	Oven temperature setting lower than 460°C.	1	Limit the change of oven setting is done by line leader only	2	12							
	Die in oven more than 1 day.	Die bearing easily wear and tear	6	36	Too early put dies into the oven.	6	Record time of die put into the oven and extrusion time	4	144	Install screen monitor to shows the time in and duration in oven. Alert by colour any less and over duration.	Extrusion HOD & Maintenance (June 2020)	6	2	12	4	48
Billet Loading	Billet lot no used not tally with extrusion record	Profile have surface defect	6	42	Mix of batch lot no of billet on loading table.	7	Visual check billet lot number, arrange billet log with lot batch no in sequence and record in Jobsheet and production record.	2	84							
	Wrong billet alloy used	Profile have surface defect	6	24	Mix of unfinish-used of another billet alloy on loading table.	3	Visual check billet lot number and record in Jobsheet and production record.	4	96							
		Die bearing damaged	7													
		Uneven anodising colour	6													
		Profile too hard for customer to assembly	8													
Extrusion	Billet temperature lower than 440°C	Hardness lower than 11wbs after ageing	8	36	Oven temperature setting lower than 440°C.	4	Using thermometer to check first billet temperature.	5	180	Increase temperature checking frequency for 2nd billet and every 10th billet.	Extrusion HOD	9	2	18	3	54
		Mechanical properties failed after ageing	9													
	Exit temperature lower than 505°C	Hardness lower than 11wbs after ageing	8	36	Billet temperature lower than 440°C.	4	Using thermometer to check first extruded exit temperature	6	216	Increase temperature checking frequency for 2nd billet and every 10th billet.	Extrusion HOD	9	2	18	3	54
		Mechanical properties failed after ageing	9													
	Billet length wrong setting	Too many unnecessary scrap (low recovery)	6	18	Wrong calculation	3	Visual inspection at cutting table	3	54							
	Bubbles	Too many unnecessary scrap at billet joint (low recovery)	8	40	burping cycle setting	5	Visual inspection at press tunnel	3	120	Annual checking burping cycle setting and strictly checking and segregate after cutting.	Extrusion HOD & Maintenance	8	4	32	3	96
	Dimension not within spec	Failure to assembly at customer site	8	48	Using wrong or unqualified die	6	Dimensions check by calliper at press	2	96	IPQC to check dimension for 2nd billet and every 10th billet. Feedback to Dowell if found dimension out.	IPQC	8	5	40	2	80
		Dimensional rejection at Fabrication/Outgoing	8													
	Pick-up on surface	Cosmetic rejection at Fabrication	7	21	Using local billet	3	Visual inspection at press tunnel	3	63							

Figure 14.
Extrusion FMEA.

4.3 Preventive maintenance

The extrusion process is a heavily mechanical process involving the extrusion press machine and other accessories that construct the whole extrusion line. Poor preventive maintenance will cause extensive unscheduled downtime and reduce productivity. Bad condition of machine and accessories will also cause quality problem. The extrusion run-out table is mainly consisting of roller and belt conveyor. The fibre material of the roller and belt are subject to wear and tear after some time. The aluminium profiles are inevitably touching with the roller and belt during material handling process. Poor surface of the fibre material will cause scratches on the aluminium profiles. Therefore, the roller and conveyor belt must be changed whenever the surface has deteriorated. The extrusion press consists of many hydraulic cylinders as its major mechanical force. Oil leakage is a major issue and weekly inspection must be done on the main cylinder, side cylinders, container cylinders and shear cylinder. The line filters for servo control and oil cooler must be changed periodically before they are clogged. The shear blade of extrusion press that cut the butt end and the shear blade of billet preheat oven must be replaced when the blades are blunt. The water volume and pressure of the cooling chamber (usually mist spray at high pressure) must be inspected to be working in good condition because effective cooling is important to T6 tempered products. Rough saw cutting station must be maintained to be free of chips sticking to the conveyor belt and roller to prevent scratches. Operators are asked to always blow the chip off and do thorough cleaning every end of shift. **Figure 15** below shows the good vs. bad conditions in a typical extrusion press.

4.4 Material handling

Aluminium surface is fragile and susceptible to scratched and dented damages when handled improperly. The extruded profile is long ranging from 20 to 40 m.

Figure 15.
Good vs. bad conveyor.

Figure 16.
Good vs. bad materials stacking on conveyor table.

Due to asymmetric contraction after cooling the profile tends to be in banana shape on the cooling table. Such a long profile needs two operators to handle at both ends. However, sometimes the operator handles it alone at one end, he will flip the profile and drag it on the belt conveyor surface, and this inclines to cause scratches and dented marks on the profile. When the operators are doing stretching, they have to carry the profile to the stretcher clamping platform, they are not able to lift up the profile from the table but to drag it for positioning and damages will be incurred if the handling is rough. Therefore, the supervisor and line leader are instructed to train their operators to do proper material handling on the conveyor table. The profiles must also align with proper distance in between them on the conveyor table to prevent knocking each other. **Figure 16** shows an example of the improper material handling and arrangement on the conveyor table that prone to cause damages.

4.5 8D report

Defects detected in the shipment lots to customer will trigger "general customer complain report" (GCCR) if a single profile records a defect of 2% out of the total delivered quantity. QA department is responsible to answer the GCCR in 7 days after confirmation of the defects by sales department. The QA engineer has to investigate the complaint by requesting physical sample from customer (for testing purpose) or high-definition pictures. Based on the basic information of sale order no and delivery no, the QA engineer will extract data from SAP system to obtain the information of delivery date, manufacturing date (extrusion, anodizing, powder coating, fabrication or packing), alloy type, surface finishing specification, aging report and QC inspection report. The QA engineer will then do root cause analysis and fill in the 8D report in the GCCR reply form. The 8D report consists of 8 sections of team members, problem description, containment action, define the root cause, implement the corrective actions, implement the horizontal corrective actions, and verify effectiveness

of actions and preventive actions for recurrence. Sometimes, the company will send QC inspectors to customer's premise to do sorting or rework. If the reject quantity is huge, materials might be sent back for rework or scrap. The practice of 8D report will be recorded to the drawing file of the profile to alert the production and QC inspector of the complaint so as to take precaution in manufacturing and inspection process in

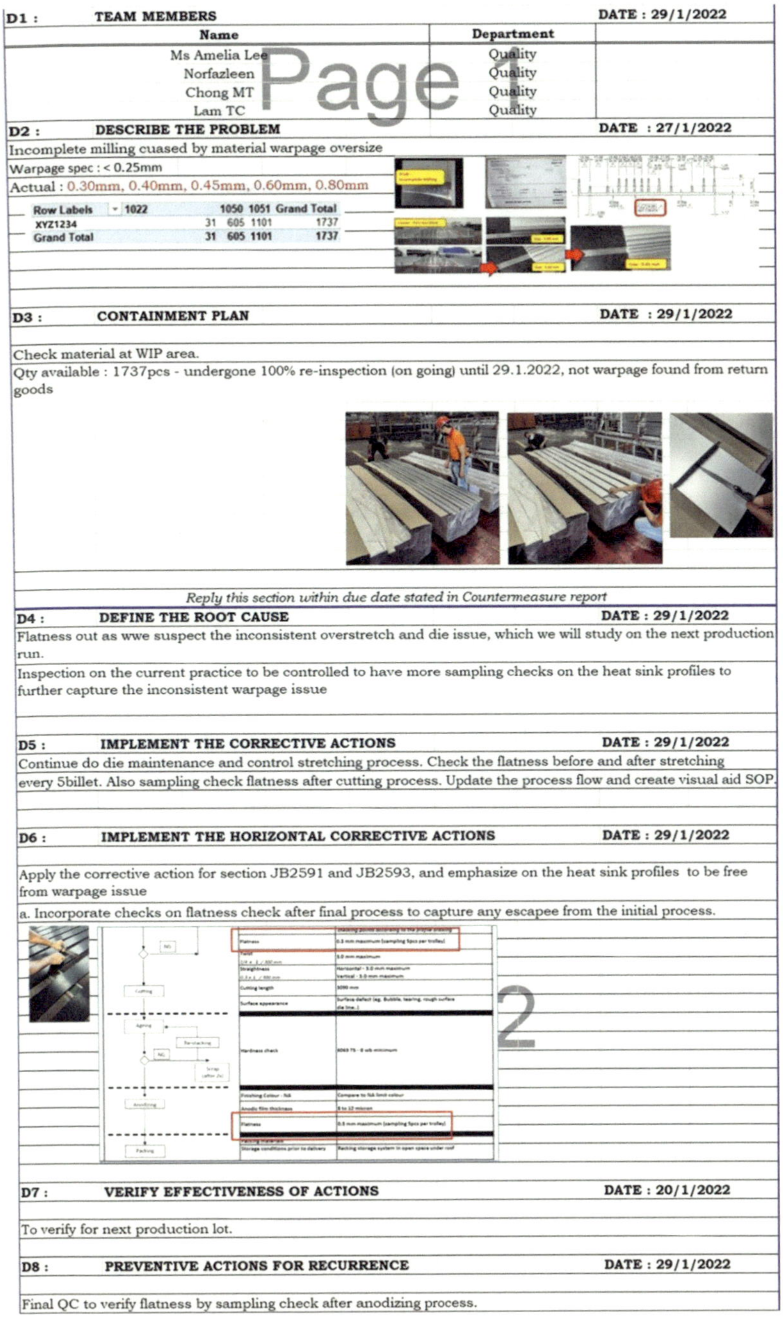

D1 : TEAM MEMBERS — DATE : 29/1/2022

Name	Department
Ms Amelia Lee	Quality
Norfazleen	Quality
Chong MT	Quality
Lam TC	Quality

Page 1

D2 : DESCRIBE THE PROBLEM — DATE : 27/1/2022

Incomplete milling cuased by material warpage oversize

Warpage spec : < 0.25mm

Actual : 0.30mm, 0.40mm, 0.45mm, 0.60mm, 0.80mm

Row Labels	1022	1050	1051	Grand Total
XYZ1234		31	605 1101	1737
Grand Total		31	605 1101	1737

D3 : CONTAINMENT PLAN — DATE : 29/1/2022

Check material at WIP area.

Qty available : 1737pcs - undergone 100% re-inspection (on going) until 29.1.2022, not warpage found from return goods

Reply this section within due date stated in Countermeasure report

D4 : DEFINE THE ROOT CAUSE — DATE : 29/1/2022

Flatness out as wwe suspect the inconsistent overstretch and die issue, which we will study on the next production run.

Inspection on the current practice to be controlled to have more sampling checks on the heat sink profiles to further capture the inconsistent warpage issue

D5 : IMPLEMENT THE CORRECTIVE ACTIONS — DATE : 29/1/2022

Continue do die maintenance and control stretching process. Check the flatness before and after stretching every 5billet. Also sampling check flatness after cutting process. Update the process flow and create visual aid SOP.

D6 : IMPLEMENT THE HORIZONTAL CORRECTIVE ACTIONS — DATE : 29/1/2022

Apply the corrective action for section JB2591 and JB2593, and emphasize on the heat sink profiles to be free from warpage issue

a. Incorporate checks on flatness check after final process to capture any escapee from the initial process.

2

D7 : VERIFY EFFECTIVENESS OF ACTIONS — DATE : 20/1/2022

To verify for next production lot.

D8 : PREVENTIVE ACTIONS FOR RECURRENCE — DATE : 29/1/2022

Final QC to verify flatness by sampling check after anodizing process.

Figure 17.
8D report.

subsequent orders to prevent the recurrence of defects. The drawing file is uploaded in the server and the production and QC staff will always access to the latest copy of drawing to check historical customer complaint record when that particular profile is being extruded. An example of 8D report is shown in **Figure 17**.

4.6 Quality campaign

The company has launched a quality campaign with the objective to cultivate quality awareness among the employees to achieve the company's goals of quality products, excellence of services and on-time delivery as the cornerstone of promoting quality culture. Quality campaign involves top and medium management to disseminate some ideas to the workers to help them enhance their quality awareness. One of them is to "look-think-act". When a worker sees something abnormal, he has to think why it happens so? And to take action to do something right. The worker should not be ignorant of what is happening around him. He has to be always concerned about the machine is running properly, process parameters are correct, quality of products are good with minimum rejection, workplace is in proper 5S condition etc. Workers are taught to understand that quality is everyone's responsibility. Quality does not happen by chance; it is the outcome of coordinated efforts from all people involved in the process. Do not finger pointing but to work together for solutions whenever there is problem. After launching the campaign, the campaign committee conducted 5S audits at extrusion, anodizing, powder coating and fabrication departments. Research has shown that 5S is able to improve productivity and quality. Top winner of 5S of the

Figure 18.
Quality with integrity.

workstation's workers will be given certificate of appreciation, souvenir and free meal coupon. Banner and poster of quality campaign are printed and displayed at many places in office and production areas. The slogan of the quality campaign is "Quality with Integrity", **Figure 18**.

4.7 Training

Aluminium extrusion is a process that requires comprehensive knowledge to understand the critical success factors that contribute to quality and productivity. The workers are not solely required to do labour intensive job but also to understand many technical aspects of the process and product. They have to understand the importance of process parameters like billet preheat temperature, die preheat temperature, die exit temperature, container temperature, ram speed, extrusion cooling rate and stretching rate. The billet quality and its impact on product's metallurgy and mechanical properties have to be understood also. The QC inspectors must be able to identify all extrusion defects and have some basic knowledge of the possible causes of defects. The QC inspectors play an important role to verify defects and instruct production to stop and change die whenever necessary to prevent over-production of defective products. The QA department has taken initiative to write a QC Handbook for all QC inspectors to understand and practice. The handbook also serves as a training material to new staff. The company also conducted internal and external training for engineers on the topics of leadership, root cause analysis and problem solving, ISO9001 QMS Awareness, Report Writing Skills, etc.

4.8 Extrusion process parameters control

Hot extrusion is a thermal deformation process done on the solid phase of billet. The management of temperature is significant to the extrusion quality and also productivity. Different alloys of billet must be preheated according to their individual upper limit temperatures. An empirical guideline is given in **Figure 19**, quoted from ASTM B807M-06. The extrusion speed is dependent on the billet temperature. The billet temperature is controlled by setting the temperatures of heating zones in the billet preheat oven. There is a region where we aim to achieve so we can maximize the extrusion efficiency as indicated in **Figure 20**. The die exit temperature is also very important because overheating will cause tearing defect on the surface due to localized melting spot when the material passed through the bearing surface; while lower exit temperature will cause insufficient cooling slope and hence inferior

Alloy	Billet or Log Temperature	
	Upper °F	[Upper °C]
6005, 6005A, 6105	1050	565
6061, 6262	1050	565
6060, 6063, 6101, 6463, 6560	1060	570
6351, 6082,	1050	565
6066, 6070	1020	550
7004, 7005	1000	540
7029, 7046, 7116, 7129, 7146	1000	540

Figure 19.
Extrusion billet temperature high limits.

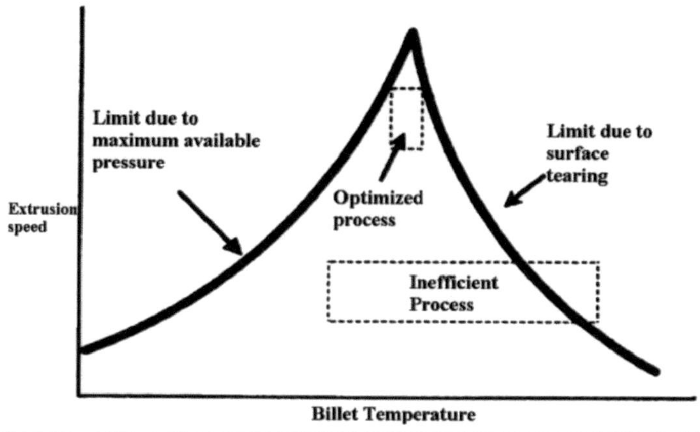

Figure 20.
Extrusion speed and billet temperature windows. Courtesy R. Peris.

Alloy	Min Die Exit °F [°C]	Min Temp Entering Quench °F [°C]	Min Cooling Rate, °F/min [°C/min]
6005, 6105	950 [510]	825 [440]	300 [165]
6005A	950 [510]	825 [440]	360 [200]
6061, 6262	930 [500]	850 [455]	600 [335]
6351, 6082	950 [510]	900 [480]	600 [335]
6060, 6063, 6101, 6463, 6560	930 [500]	825 [440]	150 [85]
6066, 6070	970 [520]	910 [490]	900 [500]
7004, 7005	750 –1000 max/ [400-540] max	725 [385]	120 [65][C]
7029, 7046, 7116, 7129, 7146	900-1000 max/ [480-540 max]	750 [400]	600 [335]

Figure 21.
Extrusion die exit temperature and cooling rate.

mechanical property. If it is T6 tempered material, minimum cooling rate has to be achieved to obtain the desired tensile and yield strength after artificial aging. The ASTM B807M-06 provides guideline on exit temperatures and cooling rates for different alloys as shown in **Figure 21**. The die exit temperature can be monitored by installing Infrared Radiation (IR) pyrometer at the machine as shown in **Figure 22**. Die exit temperature is correlated with billet preheat temperature and extrusion ram speed. These two parameters will be controlled by the operators to achieve the desired die exit temperature which is monitored in real-time by the IR pyrometer.

4.9 7S Lean workplace

The company is promoting the awareness of 7S lean workplace. 7S is defined as sort (seiri), set in order (seiton), shine (seiso), standardize (seiketsu), sustain (shitsuke), safety and spirit. It is a combination of Japanese 5S with two new elements of safety and spirit. The objectives of 7S are:

- Eliminate wastes
- Reduce space used for storage
- Streamline production
- Optimize efficiencies
- Improve safety
- Improve maintenance
- Improve quality and productivity
- Improve morale of employees

It is believed that a company that cares for its employee's safety will truly care for the product quality of its customers. An unsafe workplace will incur threat of life and stress of mind on the workers and cause disruption on their performance. Quality and productivity both suffer when employees are under stress, unsatisfied, or unable to complete their mission due to injury. But when the workplace is safe, it frees up employees to focus on their quality and their productivity. The 5S Method is a standardized process that when properly implemented creates and maintains an organized, safe, clean and efficient workplace. Improved visual controls are implemented

Figure 22.
Extrusion billet infrared temperature recorder.

as part of 5S to make any process non-conformance obvious and easily detectable [9]. Spirit refers to cultivating the interest and passion in 7S through audit, competition and reward; also to strengthen the spirit of team work among workers from top to down. The audit on safety and 5S will be conducted continuously for one week in the department and score will be given to the outcome of audit. The top scorer will be announced and presented with honorarium and certificate of appreciation.

5. Conclusions

The product quality of 6xxx series aluminium extrusion is dependent on multiple factors in the whole process stream. There are technical and management aspects that play the equivalent importance in ensuring quality and customer satisfaction. The production and QC teams must first be able to identify the extrusion defects correctly and do segregation of defects to prevent them from flowing to downstream process and wasting resources to process defective materials. PFMEA is important to guide production and maintenance team to take appropriate actions to prevent suboptimal process that will contribute to generating defects. There is sufficient knowledge base to determine the correct process parameters especially the temperature settings and control. The process must be stabilized within the controlled windows and operators must be trained to respond to anomaly in process by taking immediate actions to stop process and investigating the root cause to prevent continuous generation of defects. Historical data shows that the top five defects always dominated by scratches and dented damage which are due to improper material handling on conveyor table and trolley stacking. Intensive education and training have been provided to workers to improve their material handling. Quality can be viewed as a culture to cultivate in the workers. The company has taken initiative to launch quality campaign and 7S lean workplace campaign to promote the awareness of quality and importance of safety and housekeeping on product quality, productivity and morale. It is very important to cultivate the attitudes of continuous improvement and lifelong learning because we should not feel complacent with our current achievement and forget to make changes to cope with emerging challenges coming in our way. The extrusion industry has to grow and prosper in a sustainable manner with the commencement from optimizing its internal manufacturing process to improve quality, efficiency and reducing waste.

Author details

Ying Pio Lim* and Heng Kam Lim
PMB Aluminium Sdn Bhd, Kapar, Malaysia

*Address all correspondence to: limyp@pressmetal.com.my

References

[1] Zolotorevsky VS, Belov NA, Glazoff MV. Casting Aluminum Alloys. 1st ed. Oxford: Elsevier; 2007. p. 45

[2] Misiolek WZ, Kelly RM. Extrusion of aluminum alloys. In: ASM Handbook, Metalworking: Bulk Forming. Vol. 14A. Materials Park, Ohio: ASM International; 2005. pp. 522-527. DOI: 10.1361/asmhba0004015

[3] Pradip S. Aluminum Extrusion Technology. 1st ed. US: ASM International; 2000. pp. 1-3

[4] Uzun O, Rajendrachari S. Fundamentals of Materials Engineering—A Basic Guide. 1st ed. U.A.E: Bentham Science Publishers; 2021. DOI: 10.2174/97898114892281210101

[5] Arif AFM, Sheikh AK, Qamar SZ, Raza MK, Al-Fuhaid KM. Product defects in aluminum extrusion and its impact on operational cost. The 6th Saudi Engineering Conference, KFUPM Dharan. 2002;**5**:137-154

[6] Shamanth B, Prakash H, Subramanyam SS, Yogesh HK, Veerabhadrappa, Aravindrao MY, et al. Study of defects in aluminium extrusion process and evaluation by using quality tools. International Journal of Scientific and Engineering Research. 2021;**12**(7):355-366

[7] Hartwell J. Process Control Plan [Internet]. 2019. Available from: https://www.iqasystem.com/news/control-plan

[8] Juran. Guide to Failure Mode and Effect Analysis—FMEA [Internet]. 2018. Available from: https://www.juran.com/blog/guide-to-failure-mode-and-effect-analysis-fmea

[9] Quality-One. 5S Methodology [Internet]. 2022. Available from: https://quality-one.com/5s

Chapter 4

Low- and High-Pressure Casting Aluminum Alloys: A Review

Helder Nunes, Omid Emadinia, Manuel F. Vieira and Ana Reis

Abstract

Low- pressure casting and high-pressure casting processes are the most common liquid-based technologies used to produce aluminum components. Processing conditions such as cooling rate and pressure level greatly influence the microstructure, mechanical properties, and heat treatment response of the Al alloys produced through these casting techniques. The performance of heat treatment depends on the alloy's chemical composition and the casting condition such as the vacuum required for high-pressure casting, thus, highlighting the low-pressure casting application that does not require a vacuum. The level of pressure applied to fill the mold cavity can affect the formation of gas porosities and oxide films in the cast. Moreover, mechanical properties are influenced by the microstructure, i.e., secondary dendritic arm spacing, grain size, and the morphology of the secondary phases in the α-matrix. Thus, the current study evaluates the most current research developments performed to reduce these defects and to improve the mechanical performance of the casts produced by low- and high-pressure casting.

Keywords: aluminum alloys, low-pressure casting, high-pressure die casting, microstructure, mechanical properties

1. Introduction

Low-pressure casting (LPC) involves feeding the molten material, typically a light metal alloy such as aluminum or magnesium, into the mold cavity by applying a gas pressure onto the melt surface. This causes the melt to rise through a riser tube, placed in a crucible, and fill the mold cavity located above the furnace. This mold can be a permanent one (LPDC, low-pressure die casting) or made of sand (LPSC, low-pressure sand casting), affecting the solidification rate [1, 2]. This process can be applied to produce a vast range of components with complex geometries, such as wheels and engine crankcases [3]. Although this process requires a higher capital cost than gravity casting, it becomes more competitive by producing better-quality melts and castings with fewer defects, especially in small or medium series, which has greater production yield and allows the application of heat treatments, unlike other processes, such as high-pressure die casting [4, 5].

High-pressure casting is an established casting process for low melting temperature alloys representing about 60% of all castings used in the automotive industries.

Due to its high pressure, only permanent molds can be used, and thus it is called high-pressure die casting (HPDC). This process is characterized by a short process cycle and high productivity alongside the ability to produce parts with complex geometry, thin sections, and good surface quality. The major disadvantage of HPDC is the high cost of the equipment and dies. However, this can be compensated with production series above 5000–10,000 castings/year [6–8].

This book chapter mainly aims at comparing the process and metallurgical aspects as well as the mechanical properties of Al alloys produced by LPC and HPDC. Finally, some of the most recent developments in the casting process are discussed.

2. Metallurgy aspects of casting Al alloys

The properties of Al-Si alloy castings are significantly influenced by several microstructure features, including secondary dendrite arm spacing (SDAS), bifilms, and porosities [9].

2.1 Alloys

Aluminum alloys used in casting are often Al-Si or Al-Si-Mg, series 4xx.x and 3xx.x, respectively. The presence of silicon is critical in these alloys, since it increases the melt fluidity and decreases the coefficient of thermal expansion, facilitating casting and improving mechanical properties. The quantity of silicon added to the aluminum depends on the casting process, regarding HPDC, because of the high solidification rate, silicon contents required are between 8 and 12%, whereas, in LPSC, silicon contents between 5 and 7% are typically used. Thus, the most used and researched alloy for LPC is the A356 alloy, also known as ISO AlSi7Mg0.3 whose chemical composition can be found in **Table 1** [2, 11].

Figure 1 presents an overview of the mechanical properties obtained by various researchers of the AlSi7Mg0.3 alloy produced by LPC [4, 5, 12–18]. Most of these researchers applied the heat treatment T6 (solution heat treatment followed by water quenching and then artificial age hardening) with the aim of enhancing the mechanical properties. In this graph, Quality Index (QI) values are also represented. This index is calculated through eq. (1):

$$QI = UTS + d * log_{10}(A\%) \quad (1)$$

where ultimate tensile strength (UTS) is in MPa, A% is the elongation, and d is a constant dependent of the alloy, and for the Al-Si-Mg system *d* it is usually 150 [19].

Regarding HPDC, the most widely used alloy is AlSi9Cu3(Fe) which is normally a secondary alloy, and the chemical composition of this alloy includes some Fe as shown in **Table 2**. The effects of the presence of Fe are discussed in the next sections.

Table 3 represents some of the mechanical properties of this alloy defined in the standard NP-EN 1706 (2000) [10] and some results from different studies [20–22].

Alloy	Si	Fe	Cu	Mn	Mg	Zn	Ti	Al
AlSi7Mg0.3	6.5–7.5	<0.19	0.05	0.10	0.25–0.45	0.07	0.08–0.25	Bal.

Table 1.
Chemical composition of the AlSi7Mg0.3 (in wt.%) alloy requirements of the EN 1706 standard [10].

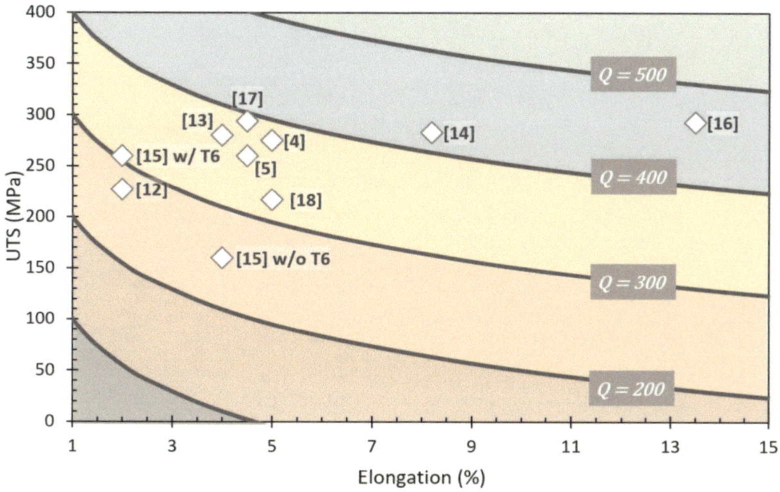

Figure 1.
State of the art of mechanical properties of the A356 with T6 alloy produced by LPC.

Alloy	Si	Fe	Cu	Mn	Mg	Cr	Ni	Zn	Pb	Sn	Ti	Al
AlSi9Cu3(Fe)	8–11	<1.3	2–4	0.55	0.05–0.55	0.15	0.55	1.2	0.35	0.25	0.25	bal

Table 2.
Chemical composition of the AlSi9Cu3 (in wt.%) alloy requirements of the EN 1706 standard [10].

	UTS (MPa)	YS (MPa)	A (%)	Hardness	QI
Standard NP EN 1706 (2000) [10]	min. 240	min. 140	> 1	min. 80 HB	min. 153
Špada et al. [20]	263 ± 3	—	1.9 ± 0.1	95 ± 2 HV	—
Cecchel et al. [21]	262 ± 3	158 ± 4	1.6 ± 0.1	~98 HV	213
Timelli et al. [22]	323	252	3.8	93 HV	387

Table 3.
Mechanical properties of the HPDC AlSi9Cu3(Fe) alloy.

The biggest difference between the properties of LPC and HPDC is the elongation values, expecting lower values of elongation due to the effect of porosities. For the AlSi9Cu3(Fe) alloy, the QI can be calculated through eq. 2 [21]:

$$QI = YS + 210 * log_{10}(A\%) + 13 \quad (2)$$

2.2 Microstructures

Figure 2 illustrates the microstructure of the AlSi7Mg0.3 alloy produced by LPDC. The as-cast microstructure (**Figure 2 a** and **b**) consist of dendrites of α-Al with the eutectic phases formed in the interdendritic spaces. In this case, the eutectic is composed of fine Si particles distributed in Al resulting from the eutectic reaction. Finally, it is also possible to identify coarse dark Mg_2Si particles (identified in the figure with red circles and arrows). As mentioned before, these types of alloys are normally

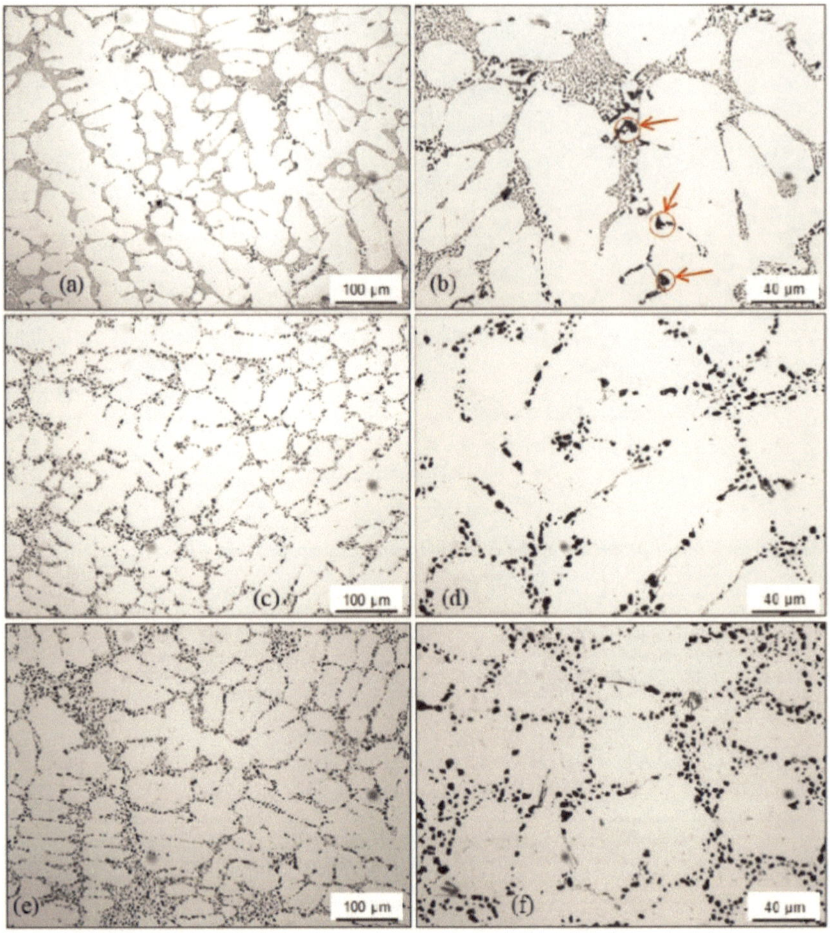

Figure 2.
Microstructure of the alloy A356.2 produced by LPDC: (a) and (b) as cast state; (c) and (d) after solubilization and quenching; (e) and (f) aged state [9].

submitted to a T6 heat treatment. The initial stages of solubilization (540°C for 6 h), quenching in water (at 40°C), and finally artificial aging (155°C for 200 min) resulted in a similar microstructure as before. However, the Si particles become coarser and more rounded, as shown in **Figure 2** f. With the increase in aging time and temperature, the UTS and yield strength (YS) tend to increase and elongation decreases due to the precipitation of very fine Mg_2Si phases, which are not possible to observe by optical microscopy [9].

Concerning the HPDC parts, the microstructure is more refined than other casting methods due to the rapid filling and fast solidification. When observing the cross section of a cylindrical casting produced by HPDC, three zones with distinct microstructures can be identified: skin layer, segregation band, and central zone (in the opposite direction of heat dissipation). When the molten Al is inside the shot sleeve, the α-Al phase starts to nucleate and grow from the walls that are commonly referred to as externally solidified crystals (ESCs). During the filling process, these crystals are forced to the center zone; thus, the solidified microstructure in this zone is composed of several coarse dendritic ESCs with sizes larger than 10 μm. Inside the die cavity, the α-Al phase continues to form however in small sizes, normally smaller than 5 μm. The high cooling rate of the melt due to the interaction with the cooler die cavity surface creates larger undercooling, promoting rapid nucleation of the α-Al phase and originating the skin layer. With ESCs continuously growing and partially interlocked during filling, liquid segregation between the central and skin zones is promoted and forms an inhomogeneous microstructure zone known as a segregation band. In this zone, the α-Al phase content is relatively low, and other phases, such as eutectic Si and intermetallic compounds, are present in larger quantities [7, 23–25].

2.2.1 Fe-rich phases

Fe is the most prejudicial contamination of Al alloys. The incorporation of these impurities occurs mainly during the recycling process and is impossible to remove by conventional methods, such as pyrometallurgy. Fe tends to react with Al to form hard and brittle intermetallic phases with a wide range of chemical formulas, sizes, and shapes. The β-Al_5FeSi is the most detrimental phase due to its plate-like shape that works as a stress concentration source and fragilizes the alloys. Recycled alloys, known as secondary aluminum alloys (SAAs), are mainly used in casting due to lower chemical restrictions needed in these processes when compared with wrought alloys. However, since the SAAs in the present recycling system is the last sink of the recycled Al alloys, a scrap surplus is expected to occur soon. Thus, it is necessary to enhance the applicability of these alloys by reducing the negative effects of the Fe-rich and obtaining SAAs with mechanical properties comparable to the primary alloys [26–28].

Even though in HPDC, the Fe can aid the ejection of the casting part from the die and prolong the die life by avoiding soldering between the two materials, and the brittle Fe-rich phases also negatively affect the mechanical properties of the alloys. In this process, the Fe-rich phases, specifically α-Al(Fe, Mn)Si can form as early as in the shot sleeve stage by nucleating in oxides from the melting furnace as Jiao et al. [24] reported in a study with AlSi10MnMg alloy. Two types of morphology were distinguished in this case, and the particles formed in the shot sleeve presented a shape like a hexahedron with a size of around 14 μm alongside the particles formed inside the cavity due to the higher cooling rate had smaller sizes, around 6 μm, with a spherical shape. In another study [7] using the same alloy, three different morphologies of the Fe-rich phase shapes were identified: polyhedral – a well-defined cube; fine compact – a claw-like shape; and Chinese script-type shape – similar to a compact skeletal structure. However, to the author's knowledge, only a few articles that study the effects of these phases on alloys produced by LPC or HPDC have been published. Thus, a more in-depth understanding of the process parameter effects on the formation of the Fe-rich phases is needed to enhance the applicability of the SAAs.

2.3 Defects

The hydrogen pickup by the melt and the formation of defects such as bifilms are the two most severe issues in Al-casting. In the LPC process, the alloy is usually heated and melted under an inert gas flux, such as argon, excessive melt oxidation, and hydrogen pickup are minimized and may provide a cleaner melt. However, the material and all equipment must be dried to remove any moisture, and the slag must be removed before casting. Since it is possible to vary the casting velocity by altering the pressure supplied to the melt, this technique is distinguished by smooth filling and good feeding capabilities. An uncontrolled filling of the mold cavity provokes a melt with high turbulence and promotes the entrapment of air. This turbulence also facilitates the molten metal to fold onto itself, which is unable to join due to the oxide layer and creates long and thin defects known as bifilms. These surface-entrained defects have been shown as the primary factor for porosity formation in LPC [6]. As a result, it is critical to ensure many processes features in the mold design and casting methods, such as preventing "waterfall" effects, which occur when molten metal falls into a depression and providing a melt velocity in the mold cavity of fewer than 0.5 m/s. These are some examples of guidelines established by Campbell to limit the melt turbulence and the number of defects [3, 7].

In the HPDC process, it is commonly verified that scrap rates of 5 to 10% due to the occurrence up to 30 specific types of defects can occur. Some of the most common defects that have a direct effect on mechanical properties are gas porosity and oxide films, similar to LPC [6, 8, 29]. Other defects can occur when further processes are applied to HPDC parts, specifically heat treatment. The high pressure associated with this process creates a high quantity of entrapped gasses in the Al. These gasses are originated from the decomposition of the die lubricants and from the entrapped air during the injection. During heat treatment, especially due to the high temperatures of the solution treatment stage, the gasses expand forming the defect known as blisters turning the piece unsuitable to use. And thus, commonly the alloys produced by HPDC are considered as not heat treatable [30].

3. LPC and HPDC methodologies and processing parameters

In this section, process cycles, working parameters, and recent advancements in LPC and HPDC are presented, as well as some effects of these aspects on the microstructure and mechanical properties of the alloys.

3.1 Equipment

The equipment used in LPC and cold chamber HPDC is shown in **Figure 3**. A furnace, a mold – which may be composed of metal or sand – and a feeder tube – which lets the metal rise from the crucible to the mold cavity – are the most common parts of the equipment required for LPC. Whereas, a shot sleeve with a hydraulic operated plunger, an intricate, and costly metal die, as well as complex systems for mold fixing, part ejection, and die cooling are the main equipment components for cold chamber HPDC [32].

3.2 Process cycle

A comparison between the two process cycles of LPC and HPDC is represented in **Figures 4** and **5**, respectively. Some similarities can be observed which are mainly the stages of filling and solidification under pressure. LPC consists initially of melting the alloy inside the furnace used for casting or by the loading of already molten Al by a ladle. When the molten is prepared for casting, the feeder tube and mold are placed on top of the furnace. By applying gas pressure, commonly with Ar, the melt rises through the feeder tube and fills the mold cavity. The pressure is maintained during the solidification of the material inside the mold. When the pressure is released, the remaining molten material falls back into the crucible. The mold is then opened when

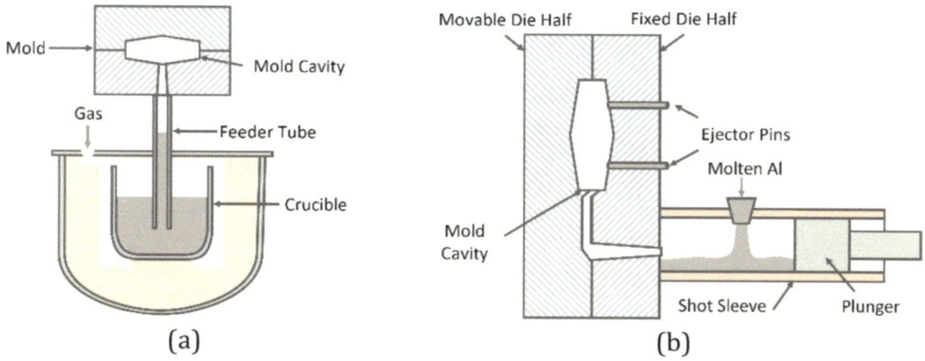

Figure 3.
General scheme: (a) LPC and (b) cold chamber HPDC (adapted from [31]).

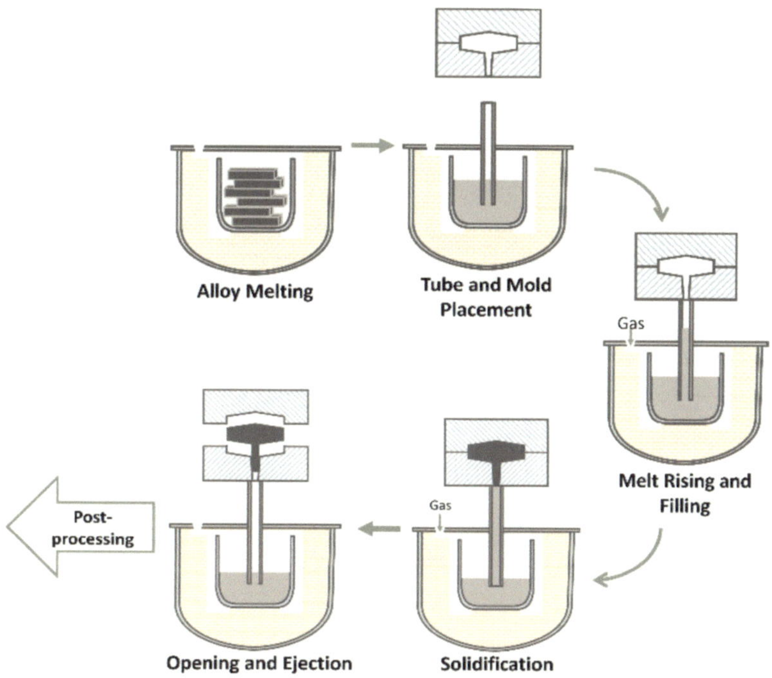

Figure 4.
General scheme of the LPC cycle [33].

dies are used, or the sand mold is destroyed by vibrations. Finally, parts can be moved to post-processing, such as heat treatment and sand-blasting [33].

The HPDC cycle mainly includes five steps, as shown in **Figure 5**: spraying and closing of the die; dosing of molten Al into the shot sleeve; injection of the melt by the application of pressures between 7 and 140 MPa through a plunger; solidification under pressure; and the opening of the die and the ejection of the part. Then the part follows to post-process, such as trimming. Mostly, the die is clamped to securely close together the two halves of the die that are already attached to the casting machine with enough force to guarantee that the die does not open during injection of the molten metal or solidification. The surface of these dies must be clean and lubricated

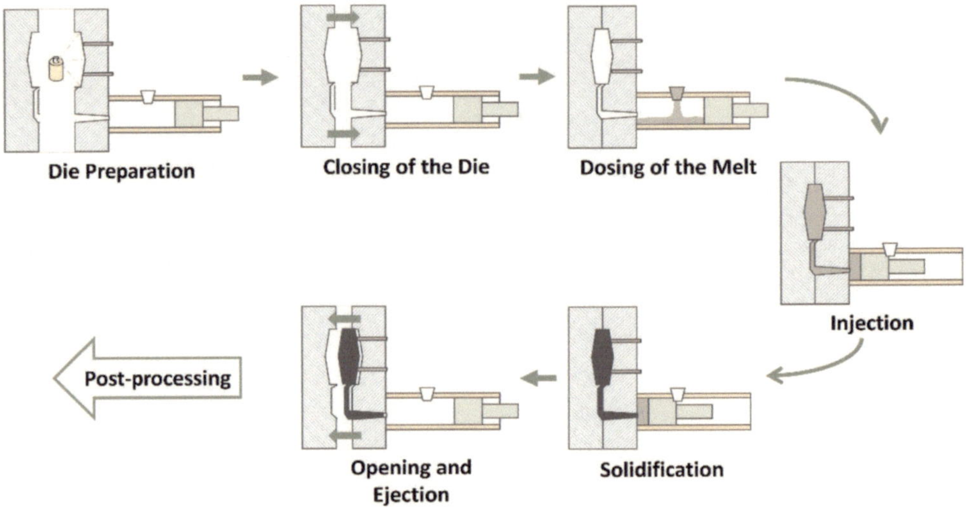

Figure 5.
General scheme of the HPDC cycle. Adapted from [31].

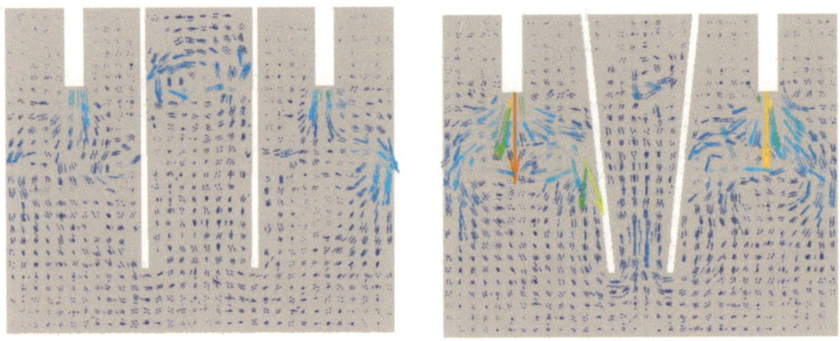

Figure 6.
Fluid velocity vector of the cylindrical riser tube (left) and the cone-shaped tube (right) [33].

to facilitate the ejection of the parts. The molten Al that was previously transferred into a chamber and part of this is then injected into the die cavity [6].

3.3 Design of die for LPC

One of the most recent studies about LPC is regarding the geometry design of the transition zone from the feeder tube to the mold cavity. While HPDC's most recent studies focus on applying a vacuum in the casting process.

Different numerical studies have been carried out to investigate the effects of geometrical parameters of the die design and the feeder tube. Yaki [33] evaluated the influence of cylindrical and cone-shaped riser tubes on liquid rising pressure and stability. The later geometry promotes a lower liquid pressure during rising and a more stable filling. This can be observed in **Figure 6**, which represents the velocity vector diagram of liquid at 10s of rising. In the case of the cylindrical tube, it is possible to observe a vortex inside the tube that increases the turbulence of the melt.

This does not occur with a conic riser tube allowing the melt to rise with more stability.

Bedel et al. [34] evaluated the impact of die geometry on filling dynamics through simulation and experimentation. Another study [35] observed that the horizontal section's geometric parameters of the furnace, the rising tube, and the mold cavity were responsible for oscillation during filling. According to specific geometric parameters studied, the section changes ratio and the section transition height impacted filling dynamics, concluding that the melt flow will be more unstable by applying greater pressure ramp and section change between the furnace and feeder tube. Thus, these researchers [34] aimed at designing an algorithm to be applied in LPC. This algorithm may be used to construct the filling system to find the proper filling pressure ramp for any complicated component. Some processes on this algorithm consist of determining the possible orientation of the parts, computation of the maximal vertical section change for each orientation, and selection of the orientation with the lowest corresponding value. It can be useful for the determination of the transition height (trans) and the actual section change (R) in a filling system to be used with a specific feeder tube. With the 3D map developed by Bedel et al. or any equivalent Lagrangian model, the maximal pressure ramp can be determined as a function of the R and trans-values. And thus, the maximum filling pressure ramp and the minimum filling system height can be determined for any component.

3.4 Vacuum-assisted HPDC

Vacuum-assisted high-pressure die casting (VHPDC) has been studied with the main purpose of reduction of entrapped air and quantities of oxide films in the cast. This is done by applying a low atmospheric pressure in the shot sleeve and cavity during injection and filling [36]. With applying vacuum, some process parameters are altered, such as the filling time, which tends to be faster. In a simulation study, Kan et al. [37], employing 500 Pa of pressure and a melt speed of 1 m/s, verified that the mold cavity under vacuum was filled in 0.95 seconds whereas the filling process took 1.2 seconds in the non-vacuum study. These results for vacuum and non-vacuum real experiments were 0.6 and 0.8 s, respectively, thus, saving about 21% of the time cycle.

In **Figure 7**, it is possible to observe the microstructures of AlSi9Cu3(Fe) alloy produced by VHPDC. These images did not show any grain size deference from similar microstructures of alloys produced without vacuum, as shown in **Figure 8** from the same study [25]. The phases α-Al, eutectic Si, and Fe-rich phases also did not

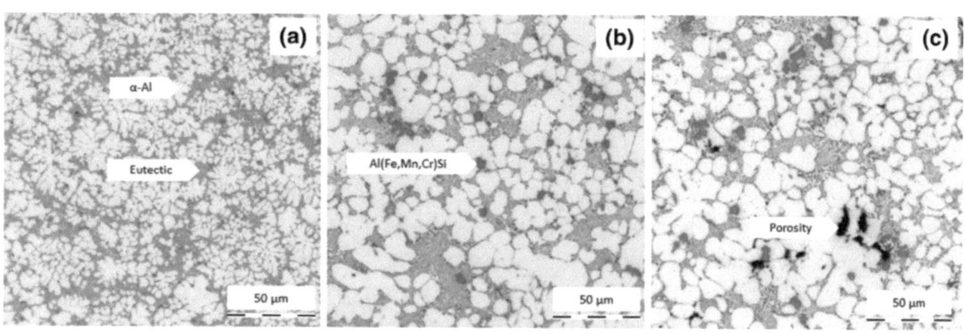

Figure 7.
Microstructure of the AlSi9Cu3(Fe) cast by VHPDC (a) skin layer; (b) central region; and (c) observation of micro-porosities [25].

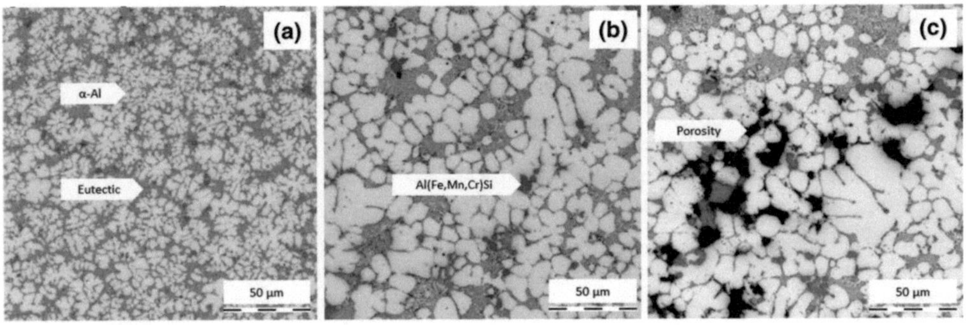

Figure 8.
Microstructure of the AlSi9Cu3(Fe) cast by HPDC (a) at the surface; (b) central region; and (c) porosities defects [25].

vary significantly with the usage of vacuum. Thus, the main differences were porosity levels and thus increased casting integrity. Reducing trapped air allows the application of heat treatment to the cast without causing the blister defects mentioned above [36]. The pores tended to be fewer and smaller, with a decrease in volumetric porosity, from 0.34 to 0.09%, and were distributed more evenly with VHPDC. This reduction in porosity enhanced the fatigue life (about 16%) of the alloy with a 4% increase in fatigue strength. The static tensile properties were improved slightly, especially UTS from 314 to 326 MPa and elongation from 2.11 to 2.81% [25]. In another study, Hu et al. [38] proved that increasing the vacuum levels can indeed improve the YS of the alloys with the reduction in pores volume.

In a study of the effect of T6 heat treatment on the alloy AlSi11MgMn, Liu et al. [39] observed a decrease in UTS, while the elongation increased for alloys produced by VHPDC. In this alloy, the microcracks formed near the large α-Fe intermetallic and not due to the eutectic Si particles. The heat treatment could change Si morphology from fibrous particles to more globular shapes. Thus, the type of fracture observed corresponded to a more ductile behavior than the fracture of the non-heat-treated alloy.

3.5 Melt treatments

In LPC, some processes should be performed on the molten alloy to guarantee good melt quality such as grain refinement and eutectic silicon modification to provide the desired mechanical properties. Grain refining of α-Al grains seeks to improve the alloys' mechanical properties, such as ultimate tensile strength and fatigue strength. It is commonly accomplished by adding B and Ti via master alloys, constituted by Al-Ti-B compounds, by the creation of Al_3Ti and/or TiB particles as nucleation agents during solidification. Although the performance of the latter type is mostly reported, a recent study revealed the better performance of Al2.2Ti1B-Mg grain refiner. This master alloy leads to the growth of an Al2.2Ti1B-Mg layer on the TiB_2 particles. Decreasing the mismatches between TiB_2 and Al promotes the nucleation of α-Al and results in a higher efficiency refining process than the other master alloys [40].

Besides, the primary goal of silicon modification is to reduce the size and form of eutectic particles to increase elongation values. The eutectic silicon modification is also done by master-alloy additions containing specific elements, such as Sr. or Na, that force the nucleation of the eutectic silicon to occur after the formation of eutectic aluminum. The presence of these elements guarantees that silicon grows between these α-Al grains and acquires a fibrous morphology, and shorter length [41]. However, the effect of these additions on defect formation is not yet thoroughly studied, with several studies showing contradictory observations. Sr additions influence the number of bifilms and the size of the pores, whereas B stimulates the formation of defects in the castings' cores. Furthermore, the Sr combines with the Al_2O_3 to generate the spinel $Sr.Al_2O_3$, facilitating the oxides to break into smaller ones [4, 42, 43]. Therefore, the additions using master alloys are a significant step in the casting process to enhance the mechanical properties by modifying the microstructures of the alloys. Moreover, alloys modified and refined have been shown to present lower porosities and higher density values than alloys without these treatments [44]. To avoid excessive porosity originating from the dissolved hydrogen, normally degassing processes are carried out before casting. Several technologies can be applied to degas the melt such as rotor degassing with argon and ultrasonic melt treatment [45, 46].

3.6 Effects of process parameters

In LPC, several parameters have a notorious effect on the properties and quality of the alloys. Some of these parameters include the mold material, the filling conditions, and the holding pressure (HP).

The type of materials used in the casting molds affects the cost and quality of the castings. Sand molds are typically less often used than permanent ones, and the use of dies allows higher productivity. However, compared to sand molds, this kind of mold requires a higher capital investment. The mechanical properties might also be impacted by the type of mold. The refined of α-Al phase, which exhibits smaller dendritic arm spacing, and eutectic Si particles in the metallic dies to induce higher values of UTS and elongation. This refinement is attributed to shorter solidification times [4].

Puga et al. [12] evaluated the effects of mold-filling parameters effects on the mechanical properties of an LPSC AlSi7Mg0.3 alloy. In this study, two different pressure-time curves were evaluated for two temperatures (650 and 700°C). The main difference in these curves is the number of ramps. While one curve only has two ramps, the other curve presents an intermediate third ramp which controls and reduces the filling velocity to smaller than 0.5 m/s. This last curve allowed a smoother filling with lower pressurized speed (Pa/s) and created a casting with fewer defects, such as porosities and oxides. And thus, the casting produced with a 3-ramp curve at 650°C presented the highest values of UTS (253 ± 9 MPa), yield strength (YS = 215 ± 5 MPa), and elongation (2.4 ± 0.2%). These authors revealed that the application of ultrasonic degassing treatment promoted the refinement of the alloys. This treatment at the lowest temperature (650°C) provoked a more intensive grain refinement and a more globular microstructure and enhanced the mechanical properties.

The LPC allows solidification to occur under a certain pressure known as holding pressure (HP). Several researchers, for example, Timelli et al. [47] and Wu et al. [16], reported a relation between HP with several characteristics such as SDAS and porosity. The local cooling rate during solidification has an impact on SDAS, which is a commonly used aspect to determine the grain size in casting alloys. Smaller SDAS values indicate a more refined microstructure, which improves some mechanical properties [48]. Timelli et al. concluded that by increasing HP from 35 to 50 kPa, the SDAS decreased from 67 to 58 μm and the porosity levels reduced from 0.3 to 0.1%. On the other hand, Wu et al. studied an even higher HP of 85 at 300 kPa. With the highest pressure, the researchers obtained SDAS values of 39 ± 6 μm (for a cooling rate of 1°C/s) and 21 ± 2 μm (for a cooling rate of 10°C/s). The density of the alloys also reached the highest values for these conditions. With the smallest SDAS and lowest porosity levels, the alloy solidified under the highest pressure and cooled faster, presenting the highest UTS value (293 ± 11 MPa) and elongation (14 ± 1%) of all the alloys studied.

It has been widely reported that the process parameters of HPDC affect the mechanical properties and microstructures of the alloys. Cho et al. [49] observed a strong proportional relationship between dendrite arm spacing and cooling rate. With the increase of the cooling rate from 15 to 100°C/sec, the dendrite arm spacing of AlSi9Cu3 and AlSi11Cu3 alloys reduced to more than half, from 12 to 5 μm and from 8 to 5 μm, respectively.

Santos et al. [50] observed no clear correlation between pressure (35 or 70 MPa) and injection temperature (579, 643, or 709°C) with the porosity of the AISi9Cu3(Fe) alloy produced by HPDC. Samples with 70 MPa have the lowest and highest values of porosity, 3 ± 1 (579°C) and 5 ± 1 (709°C). Moreover, these parameters had some effects on the microstructure of the alloys, specifically the α-Al phase. Higher temperatures promoted a refinement of this phase, while at lower temperatures the dendrite structure tends to be fragmented. The injection temperature seems to have no significant effect on mechanical properties, but the highest values of UTS were absorbed with the highest temperature (244 ± 12 and 265 ± 8 MPa, with 35 and 70 MPa injection pressure, respectively). The elongation values were constant, between 4 and 5%, in all alloys. However, no correlation between UTS and YS could be determined with these parameters. In another study of the same alloy, Obieka et al. [51] concluded that a higher pressure of about 140 MPa provoked an increase in all the mechanical properties of the alloy: UTS, YS, elongation, hardness, and impact strength. These results were mainly attributed to the refinement of the microstructure and the various phases.

4. Conclusions

This book chapter allowed for some direct comparisons between low-pressure casting and high-pressure die casting, as follows:

- Even though HPDC needs a higher initial investment, usually it is still more profitable than LPC due to its lower cycle time and higher productivity. The cycle time is one of the major differences between the processes.

- The microstructure of the most used alloys (for LPSC AlSi7Mg and HPDC AlSi9Cu3(Fe)) obtained from the process presents some similarities, such as α-Al dendrites with eutectic particles in-between. However, the HPDC provokes a refinement of the structure due to the fast filling and cooling causing small values of SDAS.

- Castings from HPDC tend to show higher quantities of defects than LPC ones, especially porosities due to the entrapment of air and turbulence during filling.

- Some of the most recent developments in both processes were analyzed. In recent studies, algorithms have been defined to establish rules for die design dies for LPC. In HPDC, applying a vacuum has been studied to improve some mechanical properties and to allow the application of heat treatments.

Acknowledgements

The authors gratefully acknowledge FCT – Portuguese Foundation for Science and Technology (2022.11466.BD), the funding of Project AM2R – Agenda Mobilizadora para a inovação empresarial do setor das Duas Rodas (C644866475-00000012) and, Hi-rEV – Recuperação do Setor de Componentes Automóveis (C644864375-00000002), cofinanced by Plano de Recuperação e Resiliência (PRR), República Portuguesa through NextGeneration EU.

Conflict of interest

The authors declare no conflict of interest.

List of abbreviations

A%	Elongation at break
ESCs	Externally solidified crystals
HP	Holding pressure
HPDC	High-pressure die casting
LPC	Low-pressure casting
LPDC	Low-pressure die casting
LPSC	Low-pressure sand casting
QI	Quality Index
R	Section change in the filling system for low-pressure casting
SAA	Secondary aluminum alloys
SDAS	Secondary dendritic arm spacing
trans	Transition height in the filling system for low-pressure casting
UTS	Ultimate tensile strength
VHPDC	Vacuum-assisted high-pressure die casting
YS	Yield strength

Author details

Helder Nunes[1,2], Omid Emadinia[1*], Manuel F. Vieira[1,2] and Ana Reis[1,2]

1 LAETA/INEGI, Institute of Science and Innovation in Mechanical and Industrial Engineering, Porto, Portugal

2 Faculty of Engineering of the University of Porto, Porto, Portugal

*Address all correspondence to: oemadinia@inegi.up.pt

References

[1] Kridli GT, Friedman PA, Boileau JM. Chapter 7 - manufacturing processes for light alloys. In: Mallick PK, editor. Materials, Design and Manufacturing for Lightweight Vehicles. Second ed. Cambridge, UK: Woodhead Publishing; 2021. pp. 267-320

[2] Committee AIH. ASM Handbook: Casting. 10th ed. Vol. 15. Ohio, USA: ASM International; 2008

[3] Sun J, Le Q, Fu L, Bai J, Tretter J, Herbold K, et al. Gas entrainment behavior of aluminum alloy engine crankcases during the low-pressure-die-casting process. Journal of Materials Processing Technology. 2019;**266**: 274-282

[4] Chiesa F, Duchesne B, Marin G, editors. Low-pressure casting of Aluminium AlSi7Mg03 (A356) in sand and permanent molds. In: 17th International Conference on Aluminium Alloys 2020 (ICAA17), 26–29 Oct 2020. France: EDP Sciences; 2020

[5] Merchaacuten M, Egizabal P, Garciacutea de Cortaacutezar M, Irazustabarrena A, Galarraga H. Development of an innovative low pressure die casting process for aluminum powertrain and structural components. Advanced Engineering Materials. 2019;**21**(6): 1800105. 6 pp

[6] Bonollo F, Gramegna N, Timelli G. High-pressure die-casting: Contradictions and challenges. Journal of Metals. 2015;**67**(5):901-908

[7] Jiao XY, Liu CF, Guo ZP, Tong GD, Ma SL, Bi Y, et al. The characterization of Fe-rich phases in a high-pressure die cast hypoeutectic aluminum-silicon alloy. Journal of Materials Science & Technology. 2020;**51**:54-62

[8] Lattanzi L, Fabrizi A, Fortini A, Merlin M, Timelli G. Effects of microstructure and casting defects on the fatigue behavior of the high-pressure die-cast AlSi9Cu3(Fe) alloy. Procedia Structural Integrity. 2017;7:505-512

[9] Santamaría JA, Sertucha J, Redondo A, Lizarralde I, Ochoa de Zabalegui E, Rodríguez P. Towards the prediction of tensile properties in automotive cast parts manufactured by LPDC with the A356.2 alloy. Metals. 2022;**12**(4):656

[10] Qualidade IP. Alumínio e ligas de alumínio: produtos vazados: composição química e características mecânicas. In: NP EN 1706 2000. Caparica: Instituto Português da Qualidade; 2000

[11] Glazoff MV, Zolotorevsky VS, Belov NA. Casting Aluminum Alloys. Amsterdam, NL: Elsevier Science; 2010

[12] Puga H, Barbosa J, Azevedo T, Ribeiro S, Alves JL. Low pressure sand casting of ultrasonically degassed AlSi7Mg0.3 alloy: Modelling and experimental validation of mould filling. Materials & Design. 2016;**94**: 384-391

[13] Kim S-W, Lee S-J, Kim D-U, Kim M-S. Experimental investigation on tensile properties and yield strength modeling of T5 heat-treated counter pressure cast A356 aluminum alloys. Metals. 2021; **11**(8):1192

[14] Lee K, Kwon Y, Lee S. Effects of eutectic silicon particles on tensile properties and fracture toughness of A356 aluminum alloys fabricated by

low-pressure-casting, casting-forging, and squeeze-casting processes. Journal of Alloys and Compounds. 2008;**461**: 532-541

[15] So TI, Jung HC, Lee CD, Shin KS. Effects of T6-treatment on the defect susceptibility of tensile strength to microporosity variation in low pressure die-cast A356 alloy. Metals and Materials International. 2015;**21**(5): 842-849

[16] Wu X, Zhang H, Ma Z, Jia L, Zhang H. Effect of holding pressure on microstructure and mechanical properties of A356 aluminum alloy. Journal of Materials Engineering and Performance. 2018;**27**(2):483-491

[17] Park CS, Kim S, Kwon Y, Lee Y, Lee J. Mechanical and corrosion properties of rheocast and low-pressure cast A356-T6 alloy. Materials Science and Engineering A-structural Materials Properties Microstructure and Processing. 2005;**391**:86-94

[18] Jia-Min H, Hai-Dong Z, Zhen-Ming C. Microstructure and properties of A356 alloy wheels fabricated by low-pressure die casting with local squeeze. Journal of Materials Engineering and Performance. 2019;**28**(4):2137-2146

[19] Lee C. Quality index of tensile property on porosity variation in A356 casting alloys upon T6 treatment. Metals and Materials International. 2021;**27**(5): 900-913

[20] Špada V, Stanić D, Brnardić I. Investigation of the mechanical properties of AlSi9Cu3(Fe)/MWCNT nanocomposites prepared by HPDC. Materiali in tehnologije. 2019;**53**:601-606

[21] Cecchel S, Panvini A, Cornacchia G. Low solution temperature heat treatment of AlSi9Cu3(Fe) high-pressure die-casting actual automotive components. Journal of Materials Engineering and Performance. 2018; **27**(8):3791-3802

[22] Timelli G, Bonollo F. The influence of Cr content on the microstructure and mechanical properties of AlSi9Cu3(Fe) die-casting alloys. Materials Science and Engineering: A. 2010;**528**(1):273-282

[23] Liu F, Zhao H, Chen B, Zheng H. Investigation on microstructure heterogeneity of the HPDC AlSiMgMnCu alloy through 3D electron microscopy. Materials & Design. 2022; **218**:110679

[24] Jiao XY, Liu CF, Guo ZP, Nishat H, Tong GD, Ma SL, et al. On the characterization of primary iron-rich phase in a high-pressure die-cast hypoeutectic Al-Si alloy. Journal of Alloys and Compounds. 2021;**862**:158580

[25] Szalva P, Orbulov IN. Influence of vacuum support on the fatigue life of AlSi9Cu3(Fe) aluminum alloy die castings. Journal of Materials Engineering and Performance. 2020; **29**(9):5685-5695

[26] Løvik AN, Modaresi R, Müller DB. Long-term strategies for increased recycling of automotive aluminum and its alloying elements. Environmental Science & Technology. 2014;**48**(8): 4257-4265

[27] Van den Eynde S, Bracquené E, Diaz-Romero D, Zaplana I, Engelen B, Duflou JR, et al. Forecasting global aluminium flows to demonstrate the need for improved sorting and recycling methods. Waste Management. 2022;**137**: 231-240

[28] Gaustad G, Olivetti E, Kirchain R. Improving aluminum recycling: A

survey of sorting and impurity removal technologies. Resources, Conservation and Recycling. 2012;**58**:79-87

[29] Karakoc C, Dizdar KC, Dispinar D. Investigation of effect of conformal cooling inserts in high-pressure die casting of AlSi9Cu3. The International Journal of Advanced Manufacturing Technology. 2022;**121**(11):7311-7323

[30] Niklas A, Orden S, Bakedano A, da Silva M, Nogués E, Fernández-Calvo AI. Effect of solution heat treatment on gas porosity and mechanical properties in a die cast step test part manufactured with a new AlSi10MnMg(Fe) secondary alloy. Materials Science and Engineering: A. 2016;**667**:376-382

[31] Ferdyn M, Piątkowski J. Influence of vacuum on adjusting parameters of high pressure die casting parts from alloy AlSi9Cu3(Fe). In: Proceedings 29th International Conference on Metallurgy and Materials, 20-22 May 2020, Brno, Czech Republic. 2020. pp. 914-918

[32] Girisha VA, Joshi MM, Kirthan LJ, Bharatish A, Hegde R. Thermal fatigue analysis of H13 steel die adopted in pressure-die-casting process. Sādhanā. 2019;**44**(6):148

[33] Yaqi L, editor. Research on design and technology of aluminium alloy conductor low pressure casting die. In: 2nd International Conference on Frontiers of Materials Synthesis and Processing, 10–11 Nov 2018. UK: IOP Publishing; 2019

[34] Bedel M, Sanitas A, El Mansori M. Geometrical effects on filling dynamics in low pressure casting of light alloys. Journal of Manufacturing Processes. 2019;**45**:194-207

[35] Sanitas A, Bedel M, El Mansori M. Experimental and numerical study of section restriction effects on filling behavior in low-pressure aluminum casting. Journal of Materials Processing Technology. 2018;**254**:124-134

[36] Niu XP, Hu BH, Pinwill I, Li H. Vacuum assisted high pressure die casting of aluminium alloys. Journal of Materials Processing Technology. 2000; **105**(1):119-127

[37] Kan M, Ipek O, Koru M. An investigation into the effect of vacuum conditions on the filling analysis of the pressure casting process. International Journal of Metalcasting. 2022;**17**: 430-446

[38] Hu C, Zhao H, Wang X, Fu J. Microstructure and properties of AlSi12Fe alloy high pressure die-castings under different vacuum levels. Vacuum. 2020;**180**:109561

[39] Liu F, Zhao H, Yang R, Sun F. Microstructure and mechanical properties of high vacuum die-cast AlSiMgMn alloys at as-cast and T6-treated conditions. Materials. 2019; **12**(13):2065

[40] Zhang Y, Ji S, Fan Z. Improvement of mechanical properties of Al-Si alloy with effective grain refinement by in-situ integrated Al2.2Ti1B-Mg refiner. Journal of Alloys and Compounds. 2017; **710**:166-171

[41] Anderson K, Weritz J, Kaufman JG. ASM Handbook, Volume 2A - Aluminum Science and Technology. Ohio, USA: ASM International; 2018

[42] Uludağ M, Çetin R, Dişpinar D, Tiryakioğlu M. On the interpretation of melt quality assessment of A356 aluminum alloy by the reduced pressure test: The Bifilm index and its physical meaning. International Journal of Metalcasting. 2018;**12**(4):853-860

[43] Uludağ M, Çetin R, Dispinar D, Tiryakioğlu M. Characterization of the effect of melt treatments on melt quality in Al-7wt %Si-Mg alloys. Metals. 2017; 7(5):157

[44] Tunçay T. The effect of modification and grain refining on the microstructure and mechanical properties of A356 alloy. Acta Physica Polonica A. 2017;**131**(1): 89-91

[45] Puga H, Barbosa J, Seabra E, Ribeiro S, Prokic M. The influence of processing parameters on the ultrasonic degassing of molten AlSi9Cu3 aluminium alloy. Materials Letters. 2009;**63**(9):806-808

[46] Schmitz C. Handbook of Aluminium Recycling (2nd Edition) - Mechanical Preparation - Metallurgical Processing - Heat Treatment. Germany: Vulkan Verlag; 2014

[47] Timelli G, Caliari D, Rakhmonov J. Influence of process parameters and Sr addition on the microstructure and casting defects of LPDC A356 alloy for engine blocks. Journal of Materials Science & Technology. 2016;**32**(6): 515-523

[48] Li Y, Liu J, Zhou H, Huang W. Study on the distribution characteristics of microstructure and mechanical properties within the cylinder head of low-pressure sand cast aluminum alloy. International Journal of Metalcasting. 2021;**16**:1252-1264

[49] Cho J-I, Kim C-W. The relationship between dendrite arm spacing and cooling rate of Al-Si casting alloys in high pressure die casting. International Journal of Metalcasting. 2014;**8**(1):49-55

[50] dos Santos SL, Antunes RA, Santos SF. Influence of injection temperature and pressure on the microstructure, mechanical and corrosion properties of a AlSiCu alloy processed by HPDC. Materials & Design. 2015;**88**:1071-1081

[51] Obiekea K, Aku S, Yawas D. Effects of pressure on the mechanical properties and microstructure of die cast aluminum A380 alloy. Journal of Minerals and Materials Characterization and Engineering. 2014;**02**:248-258

Chapter 5

Additively Manufactured High-Strength Aluminum Alloys: A Review

Fahad Zafar, Ana Reis, Manuel Vieira and Omid Emadinia

Abstract

This chapter summarizes the recent advances in additive manufacturing of high-strength aluminum alloys, the challenges of printability, and defects in their builds. It further intends to provide an overview of the state of the art by outlining potential strategies for the fabrication of bulk products using these alloys without cracking. These strategies include identifying a suitable processing window of additive manufacturing using metallic powders of conventional high-strength aluminum alloys, pre-alloying the powders, and developing advanced aluminum-based composites with reinforcements introduced either by *in situ* or *ex situ* methods. The resulting microstructures and the relationship between these alloys' microstructure and mechanical properties have been discussed. Since post-processing is inevitable in several critical applications, the chapter concludes with a brief account of post-manufacturing heat treatment processes of additively manufactured aluminum alloys.

Keywords: additive manufacturing, high strength, aluminum alloy, advanced processing, challenges, defects, advanced composites

1. Introduction

Additive manufacturing (AM) of aluminum (Al) alloys has found industrial applications and now has become a firmly established field. The number of research publications regarding Laser Powder Bed Fusion (L-PBF), one of the most popular AM processes, has shown an exponential increase during the last decade [1]. This trend is certainly not unpredictable, considering the prior wide industrial use of conventional Al-alloys due to their lightweight, high specific strength, and corrosion resistance. AM further broadens the horizon of applications for Al-alloys by its ability to produce complex geometric shapes with hollow sections for weight reductions [2]. The high-strength aluminum alloys (HSAAs) are particularly interesting in the aerospace and automotive industries. Special efforts have been directed at AM of HSAAs, and interesting advances have been made in this field [3, 4], especially in the last decade. Since the L-PBF technique has attracted the most research interest and shown promising results with HSAAs, most of the discussion and references made in this chapter will be focused on L-PBF of HSAAs, limitations in processing, strengthening mechanisms, recent achievements, defects in printed materials, and possible strategies to overcome them.

Directed energy deposition (DED) [5] and wire arc additive manufacturing [6] processes have also been utilized for the manufacturing of HSAAs. Since DED offers freedom from restriction to use a closed chamber and offers the possibility of printing large structures, a brief review of DED of HSAAs is presented in this chapter.

2. Laser powder bed fusion of aluminum alloys

Most of the foundry alloys, especially those designed for casting with near-eutectic compositions, are readily printable with negligible risk of cracking, sufficient fluidity, and minimal hot tearing susceptibility (HTS). These favorable characteristics have attracted immense research interest and led to accelerated development in AM of Al-Si alloys. But these alloys could only achieve a low-medium yield strength of <300 MPa. In contrast, some wrought alloys (2xxx, 7xxx series) can achieve far higher (~300–500 MPa) yield strength. However, these alloys have not been found readily printable by L-PBF [7, 8].

2.1 Limitations in processing

The L-PBF production of HSAAs faces challenges such as characteristic columnar microstructure eventually promoting hot cracking susceptibility (HCS) [9, 10], wide solidification range [11], solute loss due to evaporation [12], limited scanning speed to avoid cracking [13], balling, oxidation, and gas porosity.

Columnar grain growth is typically observed in L-PBF processing of Al-alloys due to the direction of the maximum thermal gradient [14]. In addition, for a certain set of L-PBF process parameters, multiple ratios of temperature gradient (G) and growth rate (R) may exist in the melt pool favoring columnar growth (either in cellular, planar, or dendritic mode). This columnar growth, more specifically the cellular or dendritic growth, leads to poor strain accommodation, and degraded liquid permeability eventually leading to high HCS [15] as shown in **Figure 1a–f**.

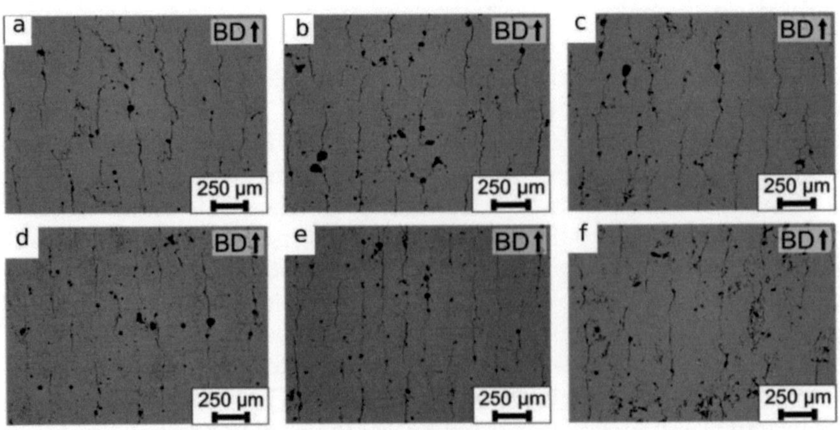

Figure 1.
(a–f) Solidification cracking observed in AlMg4.5Mn0.7 (re-printed from [16]).

HSAAs tend to have a wide solidification range (or freezing interval), which results in diminished backfill of liquid between coarse columnar crystals [17]. Solute loss occurs due to high processing temperatures during L-PBF, lower boiling points of certain alloying elements and their associated higher equilibrium vapor pressures (than that of aluminum). **Table 1** gives numerical figures for the evaporation of Zn and Mg in three different Al-alloys during L-PBF [12].

In metal deposition during additive manufacturing, liquid metal may not wet the impinging layer (or substrate) due to the surface tension of the liquid. To minimize the surface energy, the deposited liquid metal takes a spherical shape, termed *balling* (see **Figure 2a**).

Alloy	State	Zn%	Mg%	Ni%	Mn%	Cu%	Fe%	Cr%	Si%
AA2017	Before	0.21	0.72	0.009	0.57	4.0	0.40	0.016	0.56
	After	0.07	0.48	0.013	0.61	3.9	0.50	0.035	0.58
AA7020	Before	4.3	1.3	0.006	0.29	0.10	0.29	0.13	0.077
	After	3.0	1.0	0.009	0.30	0.17	0.31	0.14	0.13
AA7075	Before	5.8	2.6	0.007	0.054	1.4	0.25	0.18	0.081
	After	3.9	2.1	0.007	0.057	1.5	0.27	0.20	0.11

Table 1.
Solute element concentration before and after L-PBF [12].

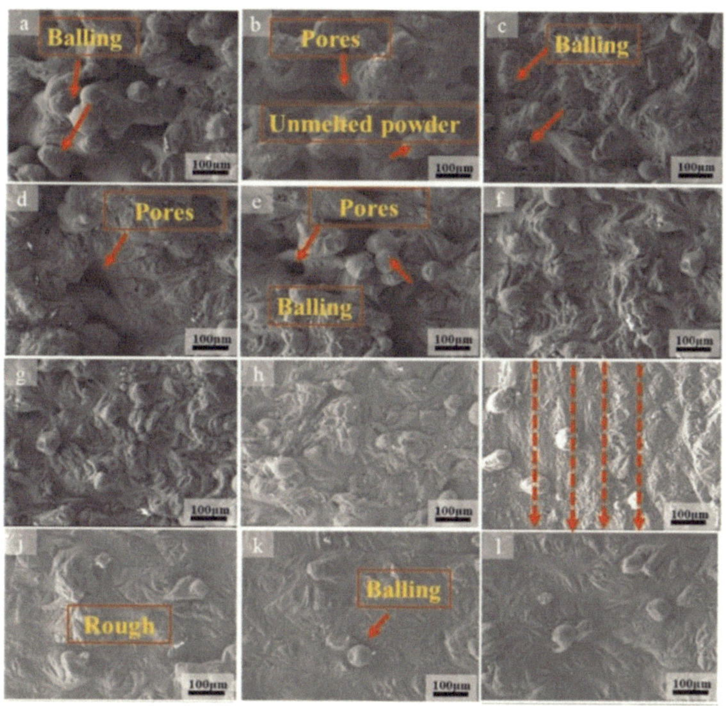

Figure 2.
(a–i) Micrographs of selective laser-melted AZ61 magnesium alloy under different laser scanning speeds showing balling and porosity defects (re-printed from [18]).

Being highly reactive toward atmospheric oxygen, aluminum alloys tend to oxidize readily by reaction with a small quantity of oxygen trapped in the air gaps between aluminum powders, which causes inferior quality in L-PBF deposited HSAAs [19].

Porosity defect has been widely reported as well as investigated in Al-alloys, and a porosity of 0.5% is generally termed acceptable in AM Al-alloys [20]. Insufficient melting of powder (or lower than the optimum volume energy density of laser) [21], moisture absorption in Al-powder from the atmosphere [21], spatter and smoke formation during AM process [22], and use of helium as inert gas for the process [23] can increase the porosity of resulting AM product.

2.2 Trends in the elimination of defects

Continued efforts have been made in the past decade to over the problems discussed above. Broadly, three main strategies have gained particular attention, showing promising results with HSAAs. These strategies are briefly listed below, and a discussion of their application and limitations will follow:

1. *Designing of new alloys,* to provide further strengthening, primarily by solid solution strengthening and/or grain boundary strengthening. Transition elements and/or rare-earth elements (e.g., Sc and Zr) have gained particular interest in this regard as dispersoids. Further strengthening through precipitation hardening may also be possible.

2. *Tailoring material or process for adaptation of existing HSAAs for AM.* This strategy seeks to enable AM of existing high-strength wrought aluminum alloys (e.g., AA2024, AA7075, and AA2219) and adapt them to AM by modifying them to diminish solidification cracking (most frequent limitation for AM of these alloys). Tailoring process parameters and defining the process window to print such existing wrought HSAAs have proved less successful. Existing high-strength foundry-grade Al-alloys can also be additively manufactured similarly.

3. *Development of advanced composite materials for AM.* This approach involves the distribution of fine (micro/nano-sized) ceramic/carbon-based reinforcement or facilitating *in situ* reaction to generate a reinforcing phase to strengthen Al-alloys.

Though designers prefer to utilize existing materials with sufficient reliable property data, it should be kept in mind that even well-established alloys present a significantly different microstructure after additive manufacturing due to rapid thermal processing. While in the previous discussion, three different strategies are presented for the elimination of defects in HSAAs, there is some overlap in these strategies to achieve an acceptable set of properties in the AM HSAAs. Moreover, there are several interdependent factors that affect the printability and the quality of the final AM product, which cannot all be discussed at length here, provided the scope of this text. **Figure 3** presents these factors, stemming either from the raw materials or the AM processing strategy.

2.3 Grain refinement strategy

The addition of nucleating agents to achieve the heterogeneous nucleation of aluminum grains upon the potent primary particles is utilized for grain refinement in

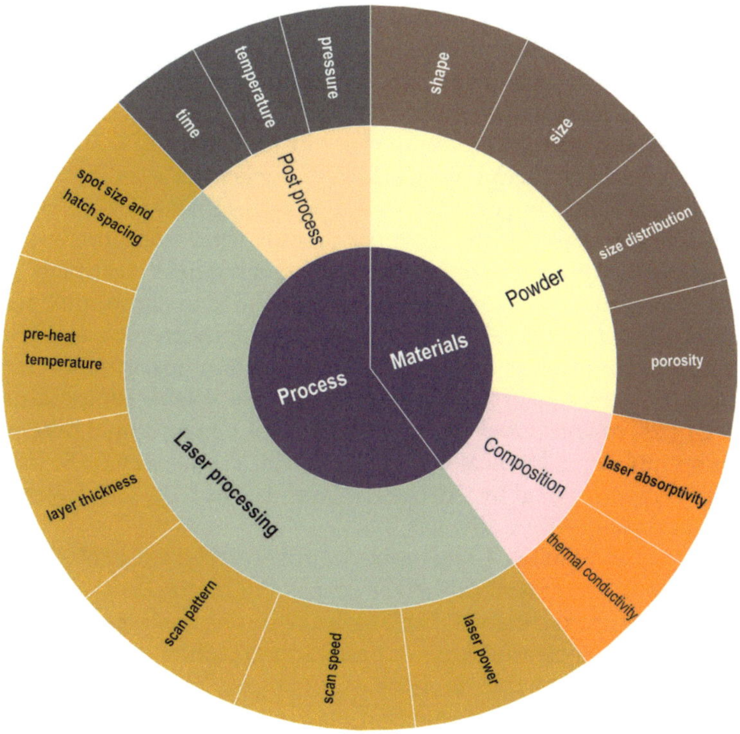

Figure 3.
Factors influencing the properties in additively manufactured HSAAs.

HSAAs. The heterogeneous nucleation promotes the formation of equiaxed grains. Such grain refinement leading to equiaxed grains is highly desirable as it offers benefits, such as reduced susceptibility to hot tearing, higher strength, lesser anisotropy, and shrinkage porosity.

A consequent reduction in the fraction of columnar grains enhances the printability of HSAAs. For equiaxed growth of a crystal, heat must dissipate from the crystal to melt ($G < 0$) [24]. In contrast, during L-PBF, heat dissipates from the melt to crystals and onward down to the substrate ($G > 0$). Thus, high enough undercooling is required to promote equiaxed grain growth. The heterogeneous nucleation diminishes the nucleation barrier by facilitating the growth of Al matrix crystals on preexisting nuclei that have a small lattice parameter misfit with that of the matrix [25]. As the growth of equiaxed grains progresses on nuclei, they impinge upon the neighboring equiaxed grains as well as the growing columnar grains, which restrict columnar growth. This phenomenon is termed as "columnar to equiaxed transition" (CET) in the solidification processing literature. The reduction in columnar grain growth also reduces the crack susceptibility in the AM HSAAs, which is a common problem faced during L-PBF of conventional wrought aluminum alloy grades.

The transition metal Scandium (Sc) and Zirconium (Zr) have best served this purpose [26, 27]. Sc provides exceptional grain refinement in aluminum alloys. The primary Sc-containing particles serve as heterogeneous nuclei, which can mitigate solidification cracking. Sc alloying imparts a significant precipitation hardening in aluminum alloys, though it is limited by the solid solubility of Sc (0.4%) in aluminum. However, rapid solidification rates in the L-PBF process enable the retention of as much as double this quantity in solution, which can be precipitation strengthened by nano-Al_3Sc precipitates during subsequent ageing treatment at 250–300°C [28]. Sc also restricts grain growth in aluminum alloys since Al_3Sc dispersoids serve to pin the grain boundaries and stabilize the grain structure [29]. These Al_3Sc particles have a small mismatch of lattice parameter with that of the aluminum matrix (0.4103 nm vs 0.4049 nm), which makes them highly effective nucleation sites for α-aluminum grains. In aluminum alloys, every 0.1% wt. Sc provides a 40–50 MPa increment in yield strength. This increase results from the precipitation strengthening by the formation of $L1_2$ coherent precipitates (Al_3Sc) during aging heat treatment [30]. Upon further addition of transition metals with low diffusivity in aluminum, such as Ti, Zr, and Hf can partially substitute Sc atoms forming precipitates such as $Al_3Sc_{1-x}Zr_x$. These resulting precipitates are highly resistant to further coarsening due to core-shell-like structure, and they offer a further advantage of high-temperature stability (aged at 325°C) [31] as compared with precipitates of conventional precipitation-hardened aluminum alloys (typically aged ~120–190°C). However, Sc has been identified as a critical raw material by European Commission [32], and alternates must be explored to offer a competitive advantage.

In an Al-Zn-Mg-Cu-Ta alloy, Ta forms *in situ* primary Al_3Ta particles and can dissolve in the second phase Al_2Cu to restrict further coarsening during heating cycles [33].

Figure 4 presents the tensile yield strengths achieved in state-of-the-art HSAAs bearing Sc and Zr, which clearly shows a possibility to achieve a yield strength higher than 500 MPa with an acceptable ductility.

Though it is worth mentioning here that multiple strengthening mechanisms may play role in strengthening, with the dominance of one or the other mechanism in the case of a particular HSAA (for further insight into strengthening mechanisms, see Ref. [34]).

2.4 Eutectic strategy and narrowing down the freezing range

Eutectic strategy is more commonly applied in L-PBF of Al-Si alloys, which facilitates sufficient backfilling of cracks. The terminal stage of solidification [35] is

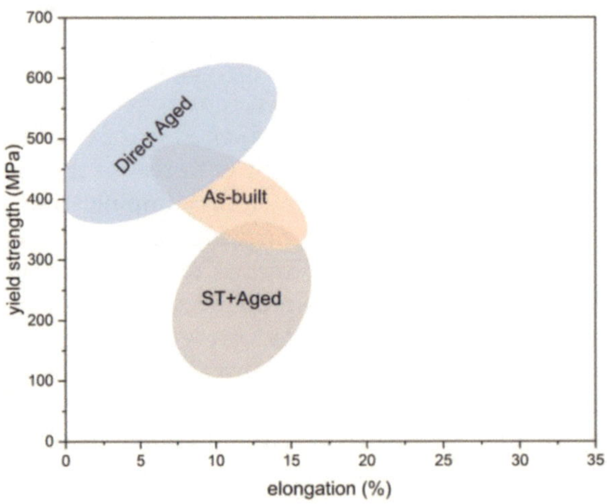

Figure 4.
Tensile yield strength (95% confidence mean) of recent Zr, Sc-strengthened HSAAs.

considered a stage with the highest hot cracking susceptibility. The conventional wrought high strength age-hardenable alloys (2xxx, 7xxx) contain alloying elements, which tend to widen their solidification range, leading to the segregation of low melting point phases during grain growth [36]. The solidification range is defined as the difference between the liquidus and solidus temperatures of the alloy. Inspired by the excellent printability of Al-Si alloys, researchers were tempted to use Si as an additive to the metal powder of existing wrought aluminum alloys. Pre-alloying with 3.74 wt.% Si reduced the freezing interval and eventually decreased the solidification cracking susceptibility in modified Al-7075 [37]. In another study, ultimate tensile strength and yield strength of 548 and 403 MPa were achieved in a newly designed alloy with Si alloying wherein numerous Al-Mg2Si fine eutectics were formed in situ upon L-PBF, which helped mitigate solidification cracking [38]. The addition of Ce in the Al matrix narrows down the freezing range. Al-3Ce-7Cu was printed successfully with 0.03% porosity and a UTS of 456 MPa in as printed condition. The alloy showed good tensile strength (UTS:186 MPa, YS:176MPa) at 250°C as well [39].

2.5 Post-processing heat treatment

High cooling rates inherent to AM process enable the achievement of unique metastable microstructures in the as-fabricated parts, which are readily transformed to equilibrium phases upon exposure to high temperatures. It is a common practice in industries to carry out a stress-relieving heat treatment to dimmish the risk of distortion or cracking due to high residual stresses after rapid thermal cycling of AM processes like L-PBF. Although in most cases, a simple heat treatment cycle serves for the intended application of AM products, some demanding applications may require a combination of this heat treatment with hot isostatic pressing.

As a general observation in the case of AM aluminum alloys, the conventional heat treatment procedures applicable for cast/wrought aluminum alloys involving solution heat treatment and ageing destroy the strengthening benefits gained through L-PBF, whereas a direct aging treatment can retain some of these benefits and still sufficiently relieves the residual stresses [40].

A typical problem faced by precipitation strengthened (e.g., Al 2xxx) aluminum alloys is posed by their low ageing temperature. While these alloys are age-hardened for strengthening between 150 and 200°C, this temperature is not high enough to relieve the residual stresses in the L-PBF parts, which need a temperature around 300°C. If stress relief is carried out at 300°C, strength in these alloys may only be regained by solution treatment followed by precipitation hardening. However, solution treatment of L-PBF precipitation hardenable alloys eradicates the beneficial effects of rapid solidification gained by rapid laser processing. Hence, it is difficult to use typical heat-treatable aluminum alloy for laser processing and gain its advantage originally inherent to artificially aged wrought aluminum alloys.

Alternatively, an alloy that can be age-hardened (or retain the strength gained during AM) at a stress-relieving temperature of ~300°C may serve as a HSAA. Such an alloy can retain the gained strength while still relieving the residual stresses with a single aging treatment (direct aging). The addition of a slow diffusion element (like Sc/Zr) to conventionally non-heat treatable Al-alloys of 3xxx and 5xxx series has induced remarkable strengthening since they tend to age-harden at much higher temperatures (≥300°C) (see **Table 2**).

Figure 5 presents the alloying element(s)-wise tensile yield strength (YS) and elongation% of some selected HSAAs from the research literature. It can be observed

Alloy	Year	Post-process	UTS (MPa)	YS (MPa)	el. %	P, SS, t, HS*	Ref.
Al-Mg-Sc-Zr	2020	St + aged	511	504	7.3	325, 1200, 30, 80	[41]
Al-Mn-Mg(Sc/Hf/Zr modified)	2022	As built	504	438	10.9	300, 700, 30, 105	[42]
Al-Mn-Mg(Sc/Hf/Zr modified)	2022	D-Ageing	542	487	7.4	300, 700, 30, 105	[42]
Al-Mn-Sc	2022	As built	412	395	21	350, 1600, 30, 100	[43]
Al-Mn-Sc	2022	D-Ageing	572	559	9.8	350, 1600, 30, 100	[43]
Al-Mn-Sc	2020	D-Ageing	573	571	16	370, 1000, 30, 100	[44]
Al-Mn-Mg-Sc-Zr	2021	As built	703	508	8.2	190, 700, 30, 100	[45]
Al-Mn-Mg-Sc-Zr	2021	D-Ageing	712	621	4.5	190, 700, 30, 100	[45]
Al-Mg-Si-Sc-Zr	2021	St + aged	580	385	19	160, 200, x, x	[46]

**P—power (watts), SS—scan speed (mm/s), t—layer thickness (mm), HS—hatch spacing (μm).*
D-Ageing—Direct aging, St + aged—solution treated and artificially aged.

Table 2.
Mechanical properties and process parameters of Sc, Zr alloyed HSAAs.

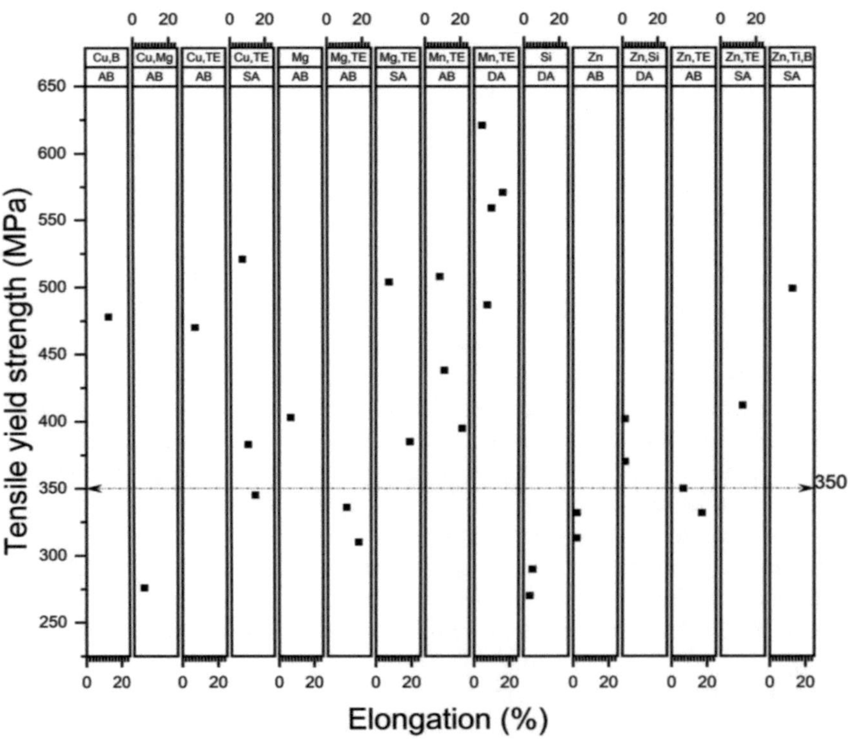

Figure 5.
Alloying element-wise tensile yield strength and elongation of selected high-strength alloys [11, 14, 38, 41–57]. TE—transition element(s), AB—as-built, SA—solution treated and aged, DA—Direct aged.

that many of these alloys demonstrate a YS higher than 350 MPa (comparable with that of conventional wrought HSAAs).

It can be inferred from **Figure 5** that a yield strength superior to 550 MPa and reasonable ductility can be achieved in Al-Mn-transition element(s) alloy after direct aging heat treatment. Without alloying of transition elements, a yield strength superior to 450 MPa has been achieved with close to 10% elongation in L-PBF using Al-Zn + Ti, B and Al-Cu + CaB_6 (composite ball-milled powders.

3. Directed energy deposition of aluminum alloys

Directed energy deposition (DED) is defined as a process in which focused thermal (laser, electron, or plasma) energy melts the feedstock material during deposition. Argon or nitrogen gas is used as a shielding gas to protect against the formation of aluminum oxide scale during thermal exposure. The most widely applied DED processes are wire arc additive manufacturing (WAAM) and laser DED (or Laser-engineered net shaping) processes.

Several studies investigated the deposition of aluminum alloys by laser-directed energy deposition (DED) [58, 59]. However, the results are not much promising, and several challenges lie ahead of DED of aluminum alloys to compete with conventional wrought alloys or their additively manufactured L-PBF counterparts.

DED of high-strength aluminum alloys faces technical challenges due to the high surface reflectivity of Aluminum, and higher laser power input is required to melt the blown aluminum powder completely [60]. This in turn leads to gas porosity by selective evaporation of lighter constituent elements such as Magnesium and Zinc [58, 61]. The high coefficient of thermal expansion (CTE) of aluminum alloys leads to shrinkage upon solidification followed by cracking or severe deformation due to intense repetitive thermal cycling during the DED process [62]. The inherent tendency of surface oxidation and moisture absorption from the atmosphere [62, 63] and poor flowability of powder, owing to low density, adversely affects the powder mass flow rate, which results in inferior quality of deposit [64].

In more recent studies on DED of high-strength Al-7075 alloy using gas-atomized powders, a crack-free, low-defect bulk material was deposited having an ultimate tensile strength of 222 $\pm$ 17 MPa and an elongation of 2%, which is significantly lower than the wrought 7075-T6 [65]. Moreover, the hardness only increased from (85 $\pm$ 4) HV0.5 to (93 $\pm$ 2) HV0.5 upon artificial aging. In another study on DED of Al-7050 alloy, hardness of a 100HV was achieved in an as-built, defect-free bulk deposit; subsequent heat treatment increased the hardness to 128 HV [61]. Due to the relative ease of deposition and wide processing window Al-Si alloys have results comparable with those of cast counterparts; however, as of now, high-strength aluminum alloys could not be deposited successfully with DED.

4. Particulate-reinforced aluminum-based composites

Aluminum matrix composites (AMCs) have already found wide applications in the aerospace and automotive industries. AM of AMCs gained particular interest due to freedom of design possibilities and further opportunities for weight reduction through modification of properties by reinforcement of Al-matrix. Either an *ex situ* or *in situ* approach is utilized for the optimization of microstructure and the resulting properties in AM. In "*ex situ*" AMCs, the reinforcements are synthesized externally and then added to the matrix, while the reinforcements are synthesized during AM process in the "*in situ*" AMCs. While *ex situ* AMCs synthesized using L-PBF have shown promising results, the limitations such as poor wettability of reinforcement by the Al-matrix, limited ability of grain refinement, and a higher likelihood of residual stress due to difference in thermophysical properties between the reinforcement and the matrix cannot be overruled [66].

Various methods are adopted for mixing powders of metallic alloys (matrix) and reinforcements. The selected method affects the resultant powder morphology and influences the laser reflectivity and the heat transfer process during L-PBF. A list of the selected methods and the factors to consider before choosing a particular process over the others is presented in **Table 3**.

Though the direct mixing process is simple, agglomeration of fine particles (nano-sized) and poor wettability are inevitable.

Ball milling is much popular due to its low cost and wide applicability to several powders. Being a nonequilibrium process, it includes a repetitive sequence of deformation, fracture, and cold welding of metal powder particles [67]. Initially, fracture occurs in brittle reinforcement particles, while cold welding dominates in the Al-matrix powder due to plastic deformation. During this deformation and cold welding, reinforcement powders are dispersed in the matrix. Since the Al-matrix gets harder following deformation and cold welding, once again fracture phase dominates until certain dynamics between cold welding and fracture ensures a stable powder size

Method	Factor	Desirability	Factor Rating
Direct mixing	cost	least	√
	applicability	Max.	√ √ √ √
	use with various vol. fractions	Max.	√ √ √ √
Ball milling	cost	Least	√ √
	applicability	Max.	√ √ √ √
	dispersion in matrix	Max.	√ √ √ √
	time	Least	√ √ √
	flowability	Max.	√
Direct mixing or ball milling + in situ reaction	cost	Least	√ √
	dispersion in matrix	Max.	√ √ √ √
	complexity	Least	√ √ √ √
In situ pre-alloying + gas-atomization	cost	Least	√ √ √ √
	flowability	Max.	√ √ √ √
	dispersion in matrix	Max.	√ √ √ √
	applicability	Max.	√ √ √ √
	complexity	Least	√ √ √ √

Table 3.
Methods for preparation of feedstock for L-PBF.

without agglomeration [68]. Ball milling induces grain refinement in the milled powders, mainly due to high energy input accompanied by severe plastic deformation into the powders. The process includes the generation of various crystal defects (dislocations, point defects), which increase the internal energy of the system(lattice) with the subsequent evolution of final grain boundaries, thus relieving the high energy. Ball milling also offers the freedom of choosing a wide size range of powders/granules as the starting material size. However, the irregular shape, rough surface, and flattening of Al-powder adversely affect the flowability of ball-milled Al-powders. Being highly ductile, Al alloys need longer duration milling cycles as compared with steels or titanium alloys. Han et al. [69] milled 4 vol.% Al_2O_3/Al powders with up to 20 h milling using different milling strategies (presented in **Table 4**).

Hence, a careful selection of a processing route for composite feedstock synthesis/production and selection of optimum processing parameters is pivotal to achieve desirable properties in AM of HSAAs.

The laser absorptivity of Al-composite powders tends to increase with the addition of reinforcements, and it increases with an increase in the amount of reinforcement in the composite. For example, the laser absorptivity of AlSi10Mg is 0.09 [70], while TiB2/AlSi10Mg composite powder has an absorptivity of 0.71 [71]. Thermal conductivity is another important thermophysical property to consider in the case of Al-matrix composites. Independent of their thermal conductivity, nano-sized reinforcement particles tend to decrease the effective thermal conductivity of the composite powders because they introduce interfacial thermal resistance and scatter the energy carriers [72].

In short, there are several considerations for the selection of an appropriate reinforcement, which are interdependent, and hence, the selection of reinforcement needs due attention. Any change in the physical and thermal property of feedstock can lead to redistribution of the thermal field and causes changes in fluid dynamics.

In a recent study, Al-2024, an age-hardenable Al-alloy powder, was modified by mechanical alloying using the ball milling technique. 0.5 w% of CaB_6 nanoparticles of 200nm (avg. size) were milled with Al-2024 powder (~63 μm) for 2h at 150 rpm, baked for moisture removal, and then printed using the L-PBF technique. A good combination of mechanical properties (UTS: 478 MPa, YS: 428 MPa, el.: 13%),

Process	Step	Powder condition
Milling for 20 h	After 4 h milling	Irregular shape, size > 100 μm
	After 8 h milling	Morphology changed, Size reduced
	After 16 h milling	Fracture took over, Size (several particles) ~20 μm, large particles observed
	After 20 h milling	Fracture continued Size range of smaller particles narrowed further (~20 μm), large particles from previous observation did not fracture
Milling for 20 h, with cyclic 10 min milling and 5 min break		some large and plate-like particles formed
Milling for 20 h, with cyclic 10 min milling and 15 min break		Fewer large particles formed a relatively more quantity of finer powder (<90 μm)

Table 4.
Ball milling parameter of 4 vol.% Al_2O_3/Al [69].

Alloy/ reinforcement (wt.%)	mfg. method/ temper	UTS, YS, el.%	Comp. powder preparation	Laser deposition conditions	Source
		MPa, MPa, %	Method	P, SS, LT, HS, Spot S.*	
Al-Cu-Mg (Al2024)/TiC (2)	L-PBF/as-built	388, 332, 10.2	BM, BPR:5:1, 130 rpm, total milling: 3 h	200, 200, x, 40, 90	[74]
Al-Cu-Mg (Al2024)/TiC (2)	L-PBF/St + aged	507, 456, 6.6	BM, BPR:5:1, 130 rpm, total milling: 3 h	200, 200, x, 40, 90	[74]
Al-Cu-Mg (Al2024)/CaB6 (1)	L-PBF/as-built	428, 478, 13	BM, BPR:5:1, 150 rpm, total milling: 3 h	200, 1000, x, 30, 100	[50]
AlSi10Mg/TiB2 + TiC (1.5 + 1.5)	L-PBF/as-built	552, 325, 12	BM, BPR:5:1, 150 rpm, total milling: 4 h	270, 1600, x, 30, 110	[75]
AlSi10Mg/TiC (3)	L-PBF/as-built	453, 267, 4.8	BM, BPR:5:1, 150 rpm, total milling: 4 h	270, 1600, x, 30, 110	[75]
AlSi10Mg/TiB2 (3)	L-PBF/as-built	361, 200, 3.8	BM, BPR:5:1, 150 rpm, total milling: 4 h	270, 1600, x, 30, 110	[75]
AlSi10Mg/TiB2 (in-situ synth.)	L-PBF/as-built	501, 320, 12.7	Pre-doped composite powder synthesis	300, x, x, x, x	[76]
Al-Cu-Mg-Ag-Ti-B (pre-doped Al-Cu/ TiB2)/TiB2 (pre-doped) (x)	L-PBF/as-built	403, 317, 10.2	Pre-doped composite powder synthesis		[77]
Al-7075/TiN (x)	WAAM/T6	503.6, x, 10.9	Powders suspended in ethanol gel and pre-placed after each layer deposition	xx	[78]
AlSi10Mg/Nbc (3)	L-PBF/as-built	393, x, 12.7	BM without balls, 71 rpm, total milling: 2 h	300, 500, 150, x, 30	[79]
AlSi10Mg/Nbc (6)	L-PBF/as-built	281, x, 9.8		300, 500, 150, x, 30	[79]

**P—power (watts) SS—scanning speed (mm/s), LT—layer thickness (μm), HS—Hatch spacing (μm), Spot S—spot size (μm).*

Table 5.
Mechanical properties and processing parameters of AMCs.

comparable with those of conventional wrought Al-2024 was achieved in as-built condition. The CaB_6 nanoparticles acted as highly effective heterogeneous nuclei due to low lattice mismatch with Al-matrix and facilitated CET, thus enabling a crack-free build [50].

Extensive research has been conducted to reveal and assess the strengthening mechanisms involved in the strengthening of composites (not discussed in this text), and different strength prediction models have been proposed. A quadratic summation strength prediction model, originally proposed by Clyne and later modified by Sanaty-Zadeh [73], can be used for estimating the strength of nano-composites.

$$\sigma_y = \sigma_{m0} + \sqrt{(\Delta\sigma_{Orowan})^2 + (\Delta\sigma_{GR})^2 + (\Delta\sigma_{load})^2 + (\Delta\sigma_{CTE})^2 + (\Delta\sigma_{Modulus})^2} \quad (1)$$

Where σ_{m0} is the yield strength of the unreinforced matrix, $\Delta\sigma_{Orowan}$ is the contribution by Orowan strengthening, $\Delta\sigma_{GR}$ is a contribution by grain size strengthening, $\Delta\sigma_{load}$ is a contribution by load-bearing strengthening, $\Delta\sigma_{CTE}$ is a contribution by dislocation density strengthening, and$\Delta\sigma_{Modulus}$ is a contribution by elastic modulus mismatch strengthening.

However, there is no consensus as to which model closely reflects reality, up till now. Since all models assume a perfect distribution of particles and bonding of interfaces, the calculated values from these models are always higher than the experimental values nevertheless showing a similar trend.

Table 5 presents the powder preparation methods, processing technique(s), mechanical properties, and laser processing conditions used for deposition in the most recent publications. These data also give a fair picture of the most recent HSAA composites and well reflect the possibility of modifying the properties (mechanical strength, wear resistance, etc.) of base aluminum alloy by the addition of appropriate reinforcement. The strengthening in such composites has reportedly been gained through multiple strengthening mechanisms, as mentioned earlier.

5. Conclusions

- A high demand exists for readily printable L-PBF high-strength light-weight parts of aluminum alloys and such alloys have been introduced.

- One of the simplest and currently feasible ways to achieve high strength is possible through using pre-alloyed Al alloy powder having Mn, Zr, Sc or Mg, Zr, and Sc, which enables to gain the benefit of high strength and stress relieving at the same time. Since scandium has already been identified as a critical raw material by European Commission, alternate grain refiners should be explored.

- The as-is printing of existing conventional wrought aluminum alloy grades with comparable mechanical properties is still challenging. However, recent progress has resulted in defect-free printing of *ex situ* reinforced wrought Al-2024 alloy with good mechanical properties, which suggests a possibility of utilizing this strategy for other aluminum alloys.

- Grain refinement through the addition of grain refiners (elements, ceramic particulates) reduces the cracking susceptibility and may also enable the material to withstand higher temperatures.

- *In situ* reinforcement approach for Al-matrix composites can achieve better dispersion and may prove more advantages than *ex situ* reinforcements. Further exploratory work is needed in this direction.

Acknowledgements

This research was funded by FEDER through programs P2020|COMPETE - Projetos em Copromoção (POCI-01-0247-FEDER-039796_LISBOA-01-0247-FEDER-039796) and P2020|COMPETE - Programas Mobilizadores (POCI-01-0247-FEDER-046100).

Author details

Fahad Zafar[1,2*], Ana Reis[1,2], Manuel Vieira[1,2] and Omid Emadinia[2]

1 Faculty of Engineering, University of Porto, Porto, Portugal

2 LAETA/INEGI–Institute of Science and Innovation in Mechanical and Industrial Engineering, Porto, Portugal

*Address all correspondence to: up202103288@edu.fe.up.pt

References

[1] Kusoglu IM, Gökce B, Barcikowski S. Research trends in laser powder bed fusion of Al alloys within the last decade. Additive Manufacturing. 2020;**36**: 101489. DOI: 10.1016/j.addma.2020.101489

[2] Olakanmi EO, Cochrane RF, Dalgarno KW. A review on selective laser sintering/melting (SLS/SLM) of aluminium alloy powders: Processing, microstructure, and properties. Progress in Materials Science. 2015;**74**:401-477. DOI: 10.1016/j.pmatsci.2015.03.002

[3] Kenevisi MS, Yu Y, Lin F. A review on additive manufacturing of Al–Cu (2xxx) aluminium alloys, processes and defects. Materials Science and Technology. 2021;**37**(9):805-829. DOI: 10.1080/02670836.2021.1958487

[4] Rometsch PA, Zhu Y, Wu X, Huang A. Review of high-strength aluminium alloys for additive manufacturing by laser powder bed fusion. Materials and Design. 2022;**219**: 110779. DOI: 10.1016/j.matdes.2022.110779

[5] Svetlizky D, Zheng B, Vyatskikh A, Das M, Bose S, Bandyopadhyay A, et al. Laser-based directed energy deposition (DED-LB) of advanced materials. Materials Science and Engineering A. 2022;**840**:142967. DOI: 10.1016/j.msea.2022.142967

[6] Vimal KEK, Naveen Srinivas M, Rajak S. Wire arc additive manufacturing of aluminium alloys: A review. Materials Today: Proceedings. 2021;**41**:1139-1145. DOI: 10.1016/j.matpr.2020.09.153

[7] Kaufmann N, Imran M, Wischeropp TM, Emmelmann C, Siddique S, Walther F. Influence of process parameters on the quality of aluminium alloy EN AW 7075 using selective laser melting (SLM). Physics Procedia. 2016;**83**:918-926. DOI: 10.1016/j.phpro.2016.08.096

[8] Karg MCH, Ahuja B, Wiesenmayer S, Kuryntsev SV, Schmidt M. Effects of process conditions on the mechanical behavior of aluminium wrought alloy EN AW-2219 (AlCu6Mn) additively manufactured by laser beam melting in powder bed. Micromachines. 2017;**8**(1): 23. DOI: 10.3390/mi8010023

[9] Elambasseril J, Benoit MJ, Zhu S, Easton MA, Lui E, Brice CA, et al. Effect of process parameters and grain refinement on hot tearing susceptibility of high strength aluminum alloy 2139 in laser powder bed fusion. Progress in Additive Manufacturing. 2022;**2022**:1-15. DOI: 10.1007/s40964-021-00259-2

[10] Hyer H, Zhou L, Park S, Huynh T, Mehta A, Thapliyal S, et al. Elimination of extraordinarily high cracking susceptibility of aluminum alloy fabricated by laser powder bed fusion. Journal of Materials Science and Technology. 2022;**103**:50-58. DOI: 10.1016/j.jmst.2021.06.023

[11] Zhang H, Zhu H, Qi T, Hu Z, Zeng X. Selective laser melting of high strength Al–Cu–Mg alloys: Processing, microstructure and mechanical properties. Materials Science and Engineering A. 2016;**656**:47-54. DOI: 10.1016/j.msea.2015.12.101

[12] Stopyra W, Gruber K, Smolina I, Kurzynowski T, Kuźnicka B. Laser powder bed fusion of AA7075 alloy: Influence of process parameters on porosity and hot cracking. Additive Manufacturing. 2020;**35**:101270. DOI: 10.1016/j.addma.2020.101270

[13] Aboulkhair NT, Simonelli M, Parry L, Ashcroft I, Tuck C, Hague R. 3D printing of aluminium alloys: Additive manufacturing of aluminium alloys using selective laser melting. Progress in Materials Science. 2019;**106**:100578. DOI: 10.1016/j.pmatsci.2019.100578

[14] Thapliyal S, Komarasamy M, Shukla S, Zhou L, Hyer H, Park S, et al. An integrated computational materials engineering-anchored closed-loop method for design of aluminum alloys for additive manufacturing. Materialia. 2020;**9**:100574. DOI: 10.1016/j.mtla.2019.100574

[15] Thapliyal S, Agrawal P, Agrawal P, Nene SS, Mishra RS, McWilliams BA, et al. Segregation engineering of grain boundaries of a metastable Fe-Mn-Co-Cr-Si high entropy alloy with laser-powder bed fusion additive manufacturing. Acta Materialia. 2021; **219**:117271 .DOI: 10.1016/j.actamat.2021.117271

[16] Böhm C, Werz M, Weihe S. Practical approach to eliminate solidification cracks by supplementing AlMg4.5Mn0.7 with AlSi10Mg powder in laser powder bed fusion. Materials. 2022;**15**(2):572. DOI: 10.3390/ma15020572

[17] Zhang X, Zheng H, Yu W. A review on solidification cracks in high-strength aluminum alloys via laser powder bed fusion. Materials Today: Proceedings. 2022. DOI: 10.1016/j.matpr.2022.09.366

[18] Liu S, Guo H. Balling behavior of selective laser melting (SLM) magnesium alloy. Materials. 2020; **13**(16):3632

[19] Altıparmak SC, Yardley VA, Shi Z, Lin J. Challenges in additive manufacturing of high-strength aluminium alloys and current developments in hybrid additive manufacturing. International Journal of Lightweight Materials and Manufacture. 2021;**4**(2):246-261. DOI: 10.1016/j.ijlmm.2020.12.004

[20] Rometsch P, Jia Q, Yang K, Wu X. 14 - Aluminum alloys for selective laser melting—towards improved performance. In: Froes F, Boyer R, editors. Additive Manufacturing for the Aerospace Industry. Amsterdam, Netherlands: Elsevier; 2019. pp. 301-325

[21] Yang KV, Rometsch P, Jarvis T, Rao J, Cao S, Davies C, et al. Porosity formation mechanisms and fatigue response in Al-Si-Mg alloys made by selective laser melting. Materials Science and Engineering A. 2018;**712**:166-174. DOI: 10.1016/j.msea.2017.11.078

[22] Anwar AB, Pham Q-C. Selective laser melting of AlSi10Mg: Effects of scan direction, part placement and inert gas flow velocity on tensile strength. Journal of Materials Processing Technology. 2017;**240**:388-396. DOI: 10.1016/j.jmatprotec.2016.10.015

[23] Wang XJ, Zhang LC, Fang MH, Sercombe TB. The effect of atmosphere on the structure and properties of a selective laser melted Al–12Si alloy. Materials Science and Engineering A. 2014;**597**:370-375. DOI: 10.1016/j.msea.2014.01.012

[24] Kurz W, Fisher DJ. Fundamentals of Solidification. Zurich, Switzerland: Trans Tech Publications Limited; 1984. p. 316

[25] Turnbull D, Vonnegut B. Nucleation catalysis. Industrial and Engineering Chemistry. 1952;**44**(6):1292-1298. DOI: 10.1021/ie50510a031

[26] Zhou L, Pan H, Hyer H, Park S, Bai Y, McWilliams B, et al. Microstructure and tensile property of a novel AlZnMgScZr alloy additively

manufactured by gas atomization and laser powder bed fusion. Scripta Materialia. 2019;**158**:24-28

[27] Liu L, Jiang J-T, Cui X-Y, Zhang B, Zhen L, Ringer SP. Correlation between precipitates evolution and mechanical properties of Al-Sc-Zr alloy with Er additions. Journal of Materials Science and Technology. 2022;**99**:61-72. DOI: 10.1016/j.jmst.2021.05.031

[28] Roder O, Wirtz T, Gysler A, Lütjering G. Fatigue properties of Al-Mg alloys with and without scandium. Materials Science and Engineering: A. 1997;**234–236**:181-184. DOI: 10.1016/S0921-5093(97)00224-4

[29] Horita Z, Furukawa M, Nemoto M, Barnes AJ, Langdon TG. Superplastic forming at high strain rates after severe plastic deformation. Acta Materialia. 2000;**48**(14):3633-3640. DOI: 10.1016/S1359-6454(00)00182-8

[30] Palm F, Leuschner R, Schubert T, Kieback B. PM Aluminiumn and Magnesium 2: Scalmalloy®= A Unique High Strength AlMgSc Type Material Concept Processed by Innovative Technologies for Aerospace Applications. European Congress and Exhibition on Powder Metallurgy European PM Conference Proceedings; 2010: The European Powder Metallurgy Association. 2010

[31] Yang KV, Shi Y, Palm F, Wu X, Rometsch P. Columnar to equiaxed transition in Al-Mg(-Sc)-Zr alloys produced by selective laser melting. Scripta Materialia. 2018;**145**:113-117. DOI: 10.1016/j.scriptamat.2017.10.021

[32] Critical raw materials: European Commission. 2020. Available from: https://single-market-economy.ec.europa.eu/sectors/raw-materials/areas-specific-interest/critical-raw-materials_en

[33] Li X, Li D, Li G, Cai Q. Microstructure, mechanical properties, aging behavior, and corrosion resistance of a laser powder bed fusion fabricated Al–Zn–Mg–Cu–Ta alloy. Materials Science and Engineering A. 2022;**832**:142364. DOI: 10.1016/j.msea.2021.142364

[34] Anderson K, Weritz J, Kaufman JG. 3.3 Heat Treatable Alloys. ASM Handbook, Volume 02A - Aluminum Science and Technology. Ohio, United States of America: ASM International

[35] Aucott L, Huang D, Dong H, Wen S, Marsden J, Rack A, et al. A three-stage mechanistic model for solidification cracking during welding of steel. Metallurgical and Materials Transactions A. 2018;**49**(5):1674-1682. DOI: 10.1007/s11661-018-4529-z

[36] Mair P, Goettgens VS, Rainer T, Weinberger N, Letofsky-Papst I, Mitsche S, et al. Laser powder bed fusion of nano-CaB6 decorated 2024 aluminum alloy. Journal of Alloys and Compounds. 2021;**863**:158714. DOI: 10.1016/j.jallcom.2021.158714

[37] Li G, Jadhav SD, Martín A, Montero-Sistiaga ML, Soete J, Sebastian MS, et al. Investigation of solidification and precipitation behavior of Si-Modified 7075 aluminum alloy fabricated by laser-based powder bed fusion. Metallurgical and Materials Transactions A. 2021;**52**(1):194-210. DOI: 10.1007/s11661-020-06073-9

[38] Yang F, Wang J, Wen T, Zhang L, Dong X, Qiu D, et al. Developing a novel high-strength Al–Mg–Zn–Si alloy for laser powder bed fusion. Materials Science and Engineering A. 2022;**851**:143636. DOI: 10.1016/j.msea.2022.143636

[39] Manca DR, Churyumov AY, Pozdniakov AV, Prosviryakov AS,

Ryabov DK, Krokhin AY, et al. Microstructure and properties of novel heat resistant Al–Ce–Cu alloy for additive manufacturing. Metals and Materials International. 2019;**25**(3): 633-640. DOI: 10.1007/s12540-018-00211-0

[40] Pereira JC, Gil E, Solaberrieta L, San Sebastián M, Bilbao Y, Rodríguez PP. Comparison of AlSi7Mg0. 6 alloy obtained by selective laser melting and investment casting processes: Microstructure and mechanical properties in as-built/as-cast and heat-treated conditions. Materials Science and Engineering: A. 2020;**778**:139124. DOI: 10.1016/j.msea.2020.139124

[41] Ma R, Peng C, Cai Z, Wang R, Zhou Z, Li X, et al. Manipulating the microstructure and tensile properties of selective laser melted Al–Mg-Sc-Zr alloy through heat treatment. Journal of Alloys and Compounds. 2020;**831**:154773. DOI: 10.1016/j.jallcom.2020.154773

[42] Li Q, Li G, Lin X, Zhu D, Jiang J, Shi S, et al. Development of a high strength Zr/Sc/Hf-modified Al-Mn-Mg alloy using Laser Powder Bed Fusion: Design of a heterogeneous microstructure incorporating synergistic multiple strengthening mechanisms. Additive Manufacturing. 2022;**57**: 102967. DOI: 10.1016/j.addma.2022.102967

[43] Bayoumy D, Kwak K, Boll T, Dietrich S, Schliephake D, Huang J, et al. Origin of non-uniform plasticity in a high-strength Al-Mn-Sc based alloy produced by laser powder bed fusion. Journal of Materials Science and Technology. 2022;**103**:121-133. DOI: 10.1016/j.jmst.2021.06.042

[44] Jia Q, Zhang F, Rometsch P, Li J, Mata J, Weyland M, et al. Precipitation kinetics, microstructure evolution and mechanical behavior of a developed Al–Mn–Sc alloy fabricated by selective laser melting. Acta Materialia. 2020;**193**: 239-251. DOI: 10.1016/j.actamat.2020.04.015

[45] Tang H, Geng Y, Bian S, Xu J, Zhang Z. An ultra-high strength over 700 MPa in Al-Mn-Mg-Sc-Zr alloy fabricated by selective laser melting. Acta Metallurgica Sinica (English Letters). 2022;**35**(3):466-474. DOI: 10.1007/s40195-021-01286-2

[46] Bi J, Lei Z, Chen Y, Chen X, Tian Z, Lu N, et al. Microstructure, tensile properties and thermal stability of AlMgSiScZr alloy printed by laser powder bed fusion. Journal of Materials Science and Technology. 2021;**69**: 200-211. DOI: 10.1016/j.jmst.2020.08.033

[47] Hu Z, Qi Y, Gao S, Nie X, Zhang H, Zhu H, et al. Aging responses of an Al-Cu alloy fabricated by selective laser melting. Additive Manufacturing. 2021; **37**:101635. DOI: 10.1016/j.addma.2020.101635

[48] Nie X, Zhang H, Zhu H, Hu Z, Ke L, Zeng X. Effect of Zr content on formability, microstructure and mechanical properties of selective laser melted Zr modified Al-4.24Cu-1.97Mg-0.56Mn alloys. Journal of Alloys and Compounds. 2018;**764**:977-986. DOI: 10.1016/j.jallcom.2018.06.032

[49] Wang Y, Lin X, Kang N, Wang Z, Liu Y, Huang W. Influence of post-heat treatment on the microstructure and mechanical properties of Al-Cu-Mg-Zr alloy manufactured by selective laser melting. Journal of Materials Science and Technology. 2022;**111**:35-48. DOI: 10.1016/j.jmst.2021.09.036

[50] Mair P, Kaserer L, Braun J, Stajkovic J, Klein C, Schimbäck D, et al.

Dependence of mechanical properties and microstructure on solidification onset temperature for Al2024–CaB6 alloys processed using laser powder bed fusion. Materials Science and Engineering A. 2022;**833**:142552. DOI: 10.1016/j.msea.2021.142552

[51] Qi Y, Zhang H, Nie X, Hu Z, Zhu H, Zeng X. A high strength Al–Li alloy produced by laser powder bed fusion: Densification, microstructure, and mechanical properties. Additive Manufacturing. 2020;**35**:101346. DOI: 10.1016/j.addma.2020.101346

[52] Michi RA, Sisco K, Bahl S, Allard LF, Wagner KB, Poplawsky JD, et al. Microstructural evolution and strengthening mechanisms in a heat-treated additively manufactured Al–Cu–Mn–Zr alloy. Materials Science and Engineering A. 2022;**840**:142928. DOI: 10.1016/j.msea.2022.142928

[53] Yang KV, Rometsch P, Davies CHJ, Huang A, Wu X. Effect of heat treatment on the microstructure and anisotropy in mechanical properties of A357 alloy produced by selective laser melting. Materials and Design. 2018;**154**:275-290. DOI: 10.1016/j.matdes.2018.05.026

[54] Casati R, Coduri M, Riccio M, Rizzi A, Vedani M. Development of a high strength Al–Zn–Si–Mg–Cu alloy for selective laser melting. Journal of Alloys and Compounds. 2019;**801**:243-253. DOI: 10.1016/j.jallcom.2019.06.123

[55] Xiao F, Wang S, Wang Y, Shu D, Zhu G, Sun B, et al. Niobium nanoparticle-enabled grain refinement of a crack-free high strength Al-Zn-Mg-Cu alloy manufactured by selective laser melting. Journal of Alloys and Compounds. 2022;**900**:163427. DOI: 10.1016/j.jallcom.2021.163427

[56] Huang B, Liu Y, Zhou Z, Cheng W, Liu X. Selective laser melting of 7075 aluminum alloy inoculated by Al–Ti–B: Grain refinement and superior mechanical properties. Vacuum. 2022;**200**:111030. DOI: 10.1016/j.vacuum.2022.111030

[57] Li L, Li R, Yuan T, Chen C, Wang M, Yuan J, et al. Microstructures and mechanical properties of Si and Zr modified Al–Zn–Mg–Cu alloy-A comparison between selective laser melting and spark plasma sintering. Journal of Alloys and Compounds. 2020;**821**:153520. DOI: 10.1016/j.jallcom.2019.153520

[58] Svetlizky D, Zheng B, Buta T, Zhou Y, Golan O, Breiman U, et al. Directed energy deposition of Al 5xxx alloy using Laser Engineered Net Shaping (LENS®). Materials and Design. 2020;**192**:108763. DOI: 10.1016/j.matdes.2020.108763

[59] Wang QZ, Lin X, Kang N, Wen XL, Cao Y, Lu JL, et al. Effect of laser additive manufacturing on the microstructure and mechanical properties of TiB2 reinforced Al-Cu matrix composite. Materials Science and Engineering A. 2022;**840**:142950. DOI: 10.1016/j.msea.2022.142950

[60] Guo C, Singh SC. Handbook of Laser Technology and Applications: Lasers: Principles and Operations (Volume One). Florida, United States of America: CRC Press; 2021

[61] Singh A, Ramakrishnan A, Dinda G. Direct laser metal deposition of Al 7050 alloy. SAE Technical Paper; 2017. Report No.: 0148-7191

[62] Zhang J, Song B, Wei Q, Bourell D, Shi Y. A review of selective laser melting of aluminum alloys: Processing, microstructure, property

and developing trends. Journal of Materials Science and Technology. 2019; **35**(2):270-284. DOI: 10.1016/j.jmst.2018.09.004

[63] Olakanmi EO. Selective laser sintering/melting (SLS/SLM) of pure Al, Al–Mg, and Al–Si powders: Effect of processing conditions and powder properties. Journal of Materials Processing Technology. 2013;**213**(8):1387-1405. DOI: 10.1016/j.jmatprotec.2013.03.009

[64] Ding Y, Muñiz-Lerma JA, Trask M, Chou S, Walker A, Brochu M. Microstructure and mechanical property considerations in additive manufacturing of aluminum alloys. MRS Bulletin. 2016;**41**(10):745-751. DOI: 10.1557/mrs.2016.214

[65] Langebeck A, Bohlen A, Rentsch R, Vollertsen F. Mechanical properties of high strength aluminum alloy EN AW-7075 additively manufactured by directed energy deposition. Meta. 2020; **10**(5):579. DOI: 10.3390/met10050579

[66] Essien UA, Vaudreuil S. In-situ metal matrix composites development for additive manufacturing: A perspective. Journal of Achievements in Materials and Manufacturing Engineering. 2022;**111**(2):78-85. DOI: 10.5604/01.3001.0015.9997

[67] Suryanarayana C. Mechanical alloying and milling. Progress in Materials Science. 2001;**46**(1-2):1-184. DOI: 10.1016/S0079-6425(99)00010-9

[68] Yu WH, Sing SL, Chua CK, Kuo CN, Tian XL. Particle-reinforced metal matrix nanocomposites fabricated by selective laser melting: A state of the art review. Progress in Materials Science. 2019;**104**:330-379. DOI: 10.1016/j.pmatsci.2019.04.006

[69] Han Q, Setchi R, Evans SL. Synthesis and characterisation of advanced ball-milled Al-Al2O3 nanocomposites for selective laser melting. Powder Technology. 2016;**297**:183-192. DOI: 10.1016/j.powtec.2016.04.015

[70] Yuan P, Gu D. Molten pool behaviour and its physical mechanism during selective laser melting of TiC/AlSi10Mg nanocomposites: Simulation and experiments. Journal of Physics D: Applied Physics. 2015;**48**(3):035303. DOI: 10.1088/0022-3727/48/3/035303

[71] Li XP, Ji G, Chen Z, Addad A, Wu Y, Wang H, et al. Selective laser melting of nano-TiB2 decorated AlSi10Mg alloy with high fracture strength and ductility. Acta Materialia. 2017;**129**:183-193. DOI: 10.1016/j.actamat.2017.02.062

[72] Ordonez-Miranda J, Yang R, Alvarado-Gil J. On the thermal conductivity of particulate nanocomposites. Applied Physics Letters. 2011;**98**(23):233111. DOI: 10.1063/1.3593387

[73] Sanaty-Zadeh A. Comparison between current models for the strength of particulate-reinforced metal matrix nanocomposites with emphasis on consideration of Hall–Petch effect. Materials Science and Engineering A. 2012;**531**:112-118. DOI: 10.1016/j.msea.2011.10.043

[74] Liu X, Liu Y, Zhou Z, Luo W, Zeng Z. Enhanced printability and strength of unweldable AA2024-based nanocomposites fabricated by laser powder bed fusion via nano-TiC-induced grain refinement. Materials Science and Engineering A. 2022;**856**:144010. DOI: 10.1016/j.msea.2022.144010

[75] Cheng W, Liu Y, Xiao X, Huang B, Zhou Z, Liu X. Microstructure and mechanical properties of a novel (TiB2+TiC)/AlSi10Mg composite prepared by selective laser melting. Materials Science

and Engineering A. 2022;**834**:142435. DOI: 10.1016/j.msea.2021.142435

[76] Xiao YK, Chen H, Bian ZY, Sun TT, Ding H, Yang Q, et al. Enhancing strength and ductility of AlSi10Mg fabricated by selective laser melting by TiB2 nanoparticles. Journal of Materials Science and Technology. 2022;**109**: 254-266. DOI: 10.1016/j.jmst.2021.08.030

[77] Li S, Cai B, Duan R, Tang L, Song Z, White D, et al. Synchrotron characterisation of ultra-fine grain TiB2/Al-Cu composite fabricated by laser powder bed fusion. Acta Metallurgica Sinica (English Letters). 2022;**35**(1):78-92. DOI: 10.1007/s40195-021-01317-y

[78] Ren X, Jiang X, Yuan T, Zhao X, Chen S. Microstructure and properties research of Al-Zn-Mg-Cu alloy with high strength and high elongation fabricated by wire arc additive manufacturing. Journal of Materials Processing Technology. 2022;**307**:117665. DOI: 10.1016/j.jmatprotec.2022.117665

[79] Raj Mohan R, Venkatraman R, Raghuraman S. Microstructure and mechanical properties of AlSi10Mg/NbC composite produced by laser-based powder bed fusion (L-PBF) process. Journal of Metals. 2022;**75**:155-166. DOI: 10.1007/s11837-022-05428-4

Section 3

Properties of Aluminum Alloys

Chapter 6

Effect of Casting Processes, Rare Earth Metals, and Sr Addition on Porosity Formation in Al-Si Cast Alloys

Ehab Samuel, Agnes M. Samuel, Victor Songmene and Fawzy H. Samuel

Abstract

The present work was carried out on A413.1cast alloy that was characterized by short freezing temperature range. Measured amounts of high purity (99.99%) rare earth metals (Ce, La, La + Ce) were added to the non-modified and Sr-modified molten metal. Three casting molds were used viz., graphite mold heated at 600°C for the purpose of obtaining solidification curves, metallic mold with three variable opening angles heated at 350°C, and a step-like metallic mold heated at 200 and 400°C. The main results are earth metals (RE) would lead to porosity formation in all molds with increase in its percentage in Sr-modified alloys. Since the maximum α-Al network formation temperature is in the range of 575–580°C, some of the RE may precipitate in the liquid state leading to blocking the flow of the liquid metal. However, considering the metal was degassed using high purity argon gas, most of the observed porosities are of shrinkage type. In addition, increasing the amount of used RE, and hence percentage of unsoluble intermetallics results in marked decrease in the alloy strength. The only observed advantage is the effectiveness of La is reducing the alloy grain size due to its low affinity to react with Ti.

Keywords: casting process, rare earth metals, porosity formation, mold type, tensile properties

1. Introduction

The rare earth elements (REE) represent the 17 metallic elements in the periodic table with high atomic numbers (57–71). This series is composed of the 15 lanthanides on the periodic table in addition to Sc (21) and Y (39), which exist in the same ores containing the oxides of the REE/RE [1–4]. Most of the RE elements were named after the places where they were found or after the names of the scientists who discovered them. The term *'rare'* was used since these types of ores were never previously reported [5]. Although there are several extraction methods to produce individual rare earth metal that have been described in the literature [6–9], little is published on their purity. It is stated that the purity of produced RE from their oxides falls between 88.5 [10] and 96.5% [11]. Also, the purity of RE depends on their prices due to the high cost of production [12].

During the past two decades, a fairly large number of articles have been published on the application of the rare earth metals La and Ce in Al-Si alloys. The authors claim that the addition of 0.2% of La or Ce produces significant modification of the eutectic Si particles [13–18], grain refining [19–21], and improvement in the alloy mechanical properties [22–24], which make RE very attractive for automotive industries. However, using high purity 99.99% La and Ce, Samuel and coworkers attributed the reported observations to the presence of tramp elements associated with the original ores and, thus, not necessarily caused by the added RE. In addition, the presence of a large volume fraction of RE-based intermetallics leads to severe deterioration of the alloy's mechanical properties [25–27]. Most of these studies were carried out on 356 alloys that were characterized by their long freezing range (6.5%Si). **Figure 1** shows a clear depression in the eutectic temperature of 356 alloys (like that reported for Sr addition [28]). However, the microstructure did not reveal a significant modification [29, 30] of the eutectic Si as claimed by the other authors.

The present work was carried out on A413.1 alloy to investigate the effect of the addition of different concentrations of La and Ce with and without Sr. This class of alloys is used for intricate thin-walled castings requiring pressure tightness, toughness, and good resistance to corrosion. The alloys have good machinability with carbide tooling. Aluminum A413. Zero is similar to Aluminum 413.0 except its iron content is lower than in A380 alloy. The 413 alloy is characterized by its short freezing range (~5°C at 0.8°C/s)). Hence, if the metal is well degassed, the only source of porosity is shrinkage porosity caused by mold geometry and temperature. To elaborate on this statement, three types of molds were used: a graphite mold (heated at 600°C), a variable angle metallic mold (heated at 350°C), and a step-like mold heated at 200°C and 400°C.

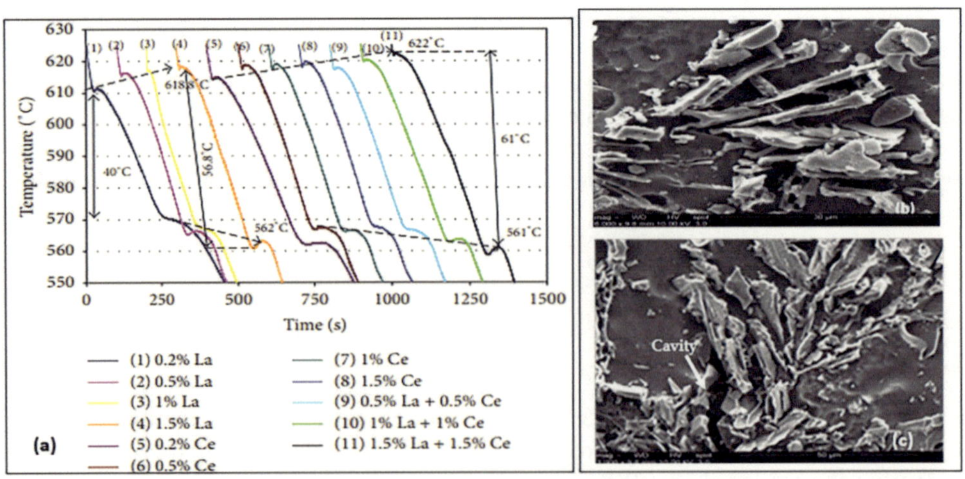

Figure 1.
Effect of addition of high purity La and Ce on the thermal behavior of 356 alloys (solidification rate of about 0.8°C/s), (b) alloy #1, (c) alloy #11. Note the absence of modification [28].

2. Experimental procedure

Table 1 lists the chemical analysis of the as-received 413 alloy. To obtain the solidification curves and to identify the main reactions and corresponding temperatures occurring during the solidification of the A413.1 alloy, thermal analysis was conducted for all the needed compositions to be prepared. The molten metal (using a 2-kg SiC crucible, heated to 750°C using an electrical furnace) for each composition was cast into a graphite mold preheated to 600°C to obtain a slow solidification rate resembling equilibrium conditions. A high-sensitivity Type-K (chromel-alumel) thermocouple was attached to the center of the graphite mold—**Figure 2(a)**; the temperature-time data was collected using a high-speed data acquisition system.

The ingots were melted in a 40 kg capacity SiC crucible using an electrical resistance furnace. The molten metal was degassed using pure, dry argon, and introduced into the melt using a graphite rotary impeller (at a speed of 130 rpm). Prior to degassing, measured amounts of Sr, La, and Ce were added. The three elements were introduced into the molten alloy in the form of Al-10%Sr, Al-20%La, and Al-20%Ce master alloys. At the end of the degassing period, the molten alloy was cast into two different molds, which provided different solidification rates:

1. A variable angle metallic mold (0∘, 5∘, and 15∘) heated at 350°C (**Figure 2(b)** and **(c)**).

2. A step-like metallic mold heated at either 200°C or 400°C (**Figures 2(d)** and **(e)**).

The melt was also poured into an ASTM B-108 permanent mold (preheated at 450°C to drive out moisture) for preparing the tensile test bar castings (gauge length of 70 mm and a crosssectional diameter of 12.7 mm). Tensile bars were solutionized at 510°C for 8 h, quenched in warm water at 60°C, and thereafter aged at 180°C for 2, 8, 20, and 50 h, followed by air cooling. All tensile bars were pulled to fracture using an MTS servo-hydraulic tensile testing machine at a strain rate of 4×10^{-4} s^{-1}. All bars were tested at 25°C, and each value represents the average of 10 tensile bars. The volume fraction of the precipitated intermetallics was determined using a JEOL JSM-6480LV scanning electron microscope coupled with an energy dispersive X-ray spectrometer (EDS) was used for examining the microstructures and semiquantitative analysis of the phases observed.

Table 2 summarizes the measured SDAS values obtained from the examined samples. Phase identification was carried out using an electron probe microanalyzer (EPMA) in conjunction with energy dispersive X-ray (EDX/EDS) analysis and wavelength dispersive spectroscopic (WDS) analysis where required, integrating a combined JEOL JXA-8900lWD/ED microanalyzer operating at 20 kV and 30 nA,

Alloy	Si	Mg	Mn	Cu	Fe	Sr	Ti	B
A413.1	11.17	0.14	0.218	0.71	0.344	0.0000	0.052	0.0003

Table 1.
Chemical composition of A413.1 alloy (% weight).

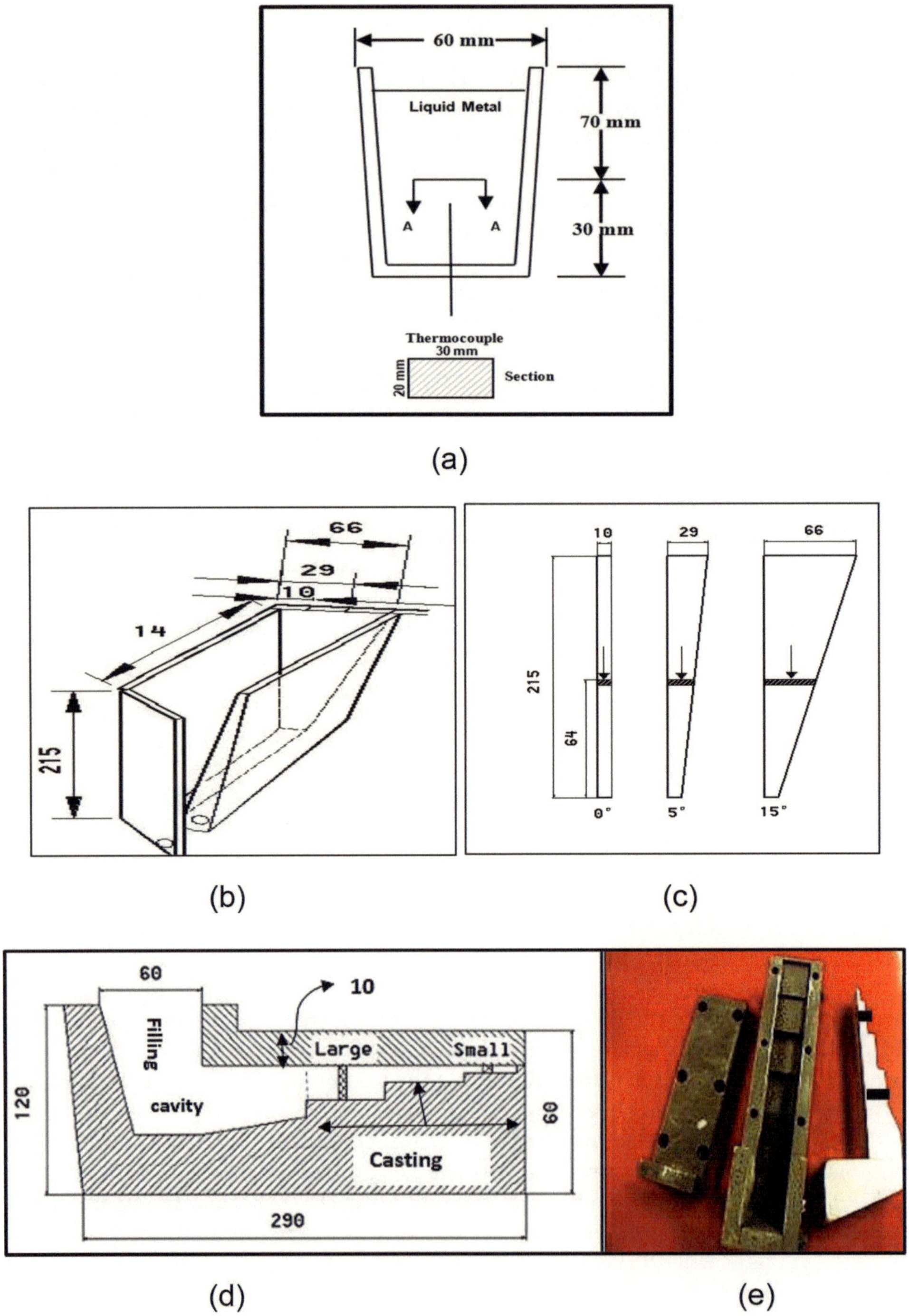

Figure 2.
(a) Schematic of the thermal analysis setup. (b) Schematic of the metallic variable mold—all dimensions are in mm. (c) Sections from actual castings—arrows indicate samples position for metallographic examination. (d) Schematic of the step-like metallic mold (small 5 mm and large 20 mm) all dimensions in mm. (e) Actual mold and casting.

Mold	Mold temp. (°C)	Mold section	SDAS (μm)	
			Average	SD
Graphite	600	Center	68.63	5.49
	200	Large	31.56	2.92
Step-like	200	Small	16.17	1.89
	400	Large	41.66	3.02
	400	Small	22.46	3.96
Variable angle	325	Large	53.62	5.7
	325	Small	25.04	3.0

Table 2.
Average secondary dendrite arm spacing of the examined A356 alloy samples.

Aimed concentration (wt%)		Actual concentration (wt%)	
La	Ce	La	Ce
0.2	0	0.187	0
0.5	0	0.384	0
1	0	0.717	0
1.5	0	0.94	0
0	0.2	0.022	0.083
0	0.5	0.237	0.308
0	1	0.067	0.82
0	1.5	0.078	1.288

Table 3.
Aimed and actual La and Ce concentrations (weight%).

where the size of the spot examined was ~2 μm. **Table 3** summarizes the actual La and Ce concentrations using spectroscopic analysis. Average concentration of added Sr was in the range 150–180 ppm.

3. Results and discussion

3.1 Thermal analysis: graphite mold

Figure 3(a) depicts the solidification curve and its first derivative obtained from the base A413.1 alloy solidified at 0.8°C/s revealing a narrow solidification range of approximately 5°C compared to 43°C reported for the commercial A356.1 alloy as listed in **Table 4**. With the addition of 180 ppm Sr, the eutectic temperature in both alloys was dropped by about 7–8°C. In the absence of Sr, with the addition of La or Ce up to 1.5% (actual concentrations are reported in **Table 3**) the maximum drop in

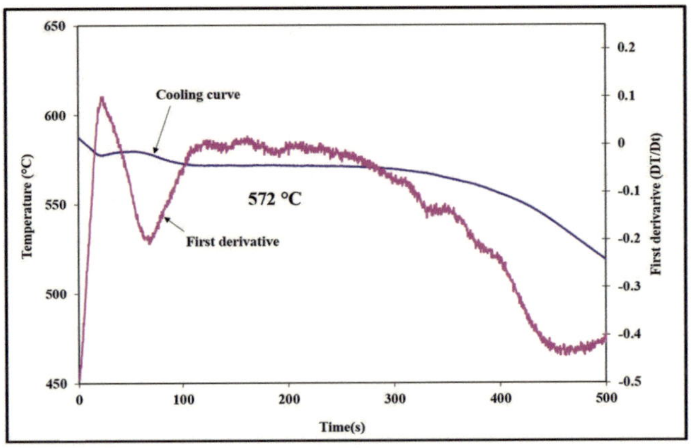

(a)

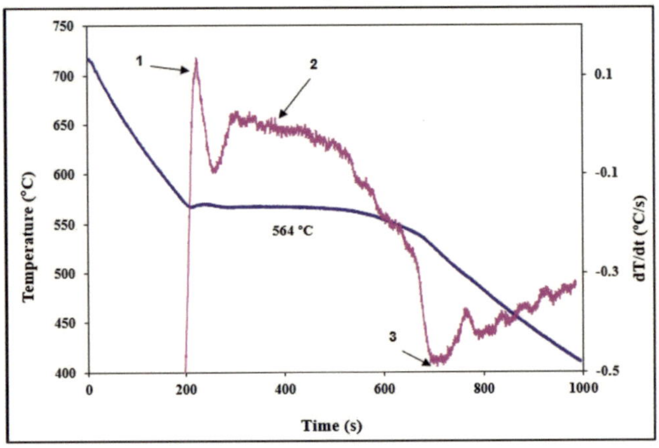

(b)

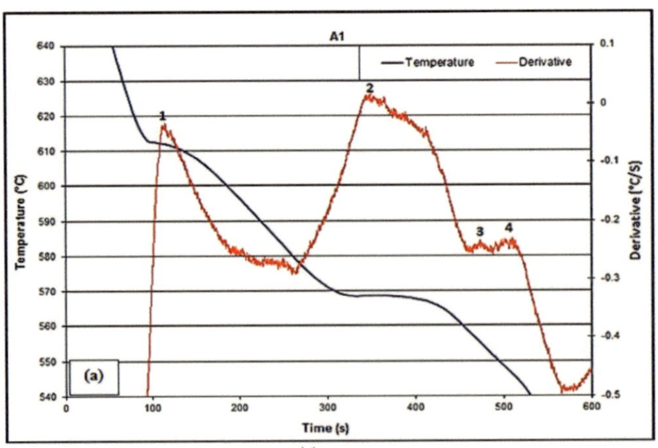

(c)

Figure 3.
Solidification curves and their first derivatives in: (a) base A413.1 alloy, (b) A413.1 alloy +180 ppm Sr, and (c) A356.1 alloy for comparison.

RE	α-Al (°C)	Al + Si (°C)	Freezing zone	ΔT (°C) increase in α-Al	ΔT (°C) decrease in Al + Si
0	577	573	4-5	0	0
Sr (180 ppm)	574	564	10	—	8
1.5 La	577	570	7	0	2
1.5 Ce	580	569	10	3	2
1.5 La +180 ppm Sr	574	565	9	—	7
1.5 Ce +180 ppm Sr	580	565	15	3	S
356 alloy	612	569	43		
356 allov+Sr	609	561	48		8

Table 4.
Effect of RE and Sr on the solidification characteristics of A413.1 alloy.

the eutectic temperature of A413.1 alloy is less than 2°C, taking into consideration that fluctuation in the solidification rate becomes more pronounced when the alloy is modified with Sr in addition to RE. According to Mahmoud et al. [28], the eutectic temperature can be expressed as:

$$T_R\,(\mathrm{oC}) = 577 - 12.5\,w\mathrm{Si}\cdot(4.59\cdot w\mathrm{Mg} + 1.37\cdot w\mathrm{Fe} + 1.65\cdot w\mathrm{Cu} + 0.35\cdot w\mathrm{Zn} + 2.54\cdot w\mathrm{Mn} + 3.52\cdot w\mathrm{Ni}) \quad (1)$$

As can be seen, Sr was not taken into consideration. Based on this equation for the present alloy, T_R for A413.1 is about 574°C as reported in **Table 4**.

According to [31], both Al-La and Al-Ce reveal a eutectic reaction at low RE concentration as depicted in **Figure 4**. From **Table 4**, the addition of La or Ce up to 1.5% seems to have a marginal effect (3–4°C) on the two main reactions in A413.1 alloy, that is, precipitation of α-Al and (Al + Si) eutectic, increasing the freezing zone to 10°C. **Figure 5(a)** exhibits the effect of gradual addition of Ce on the temperature of the (Al + Si) eutectic reaction, reaching maximum 2–3°C at 1.5%Ce, which is due to the change in the alloy composition. The addition of 180 ppm Sr was proven to be more effective in reducing the (Al + Si) eutectic reaction by about 8–10°C as shown in **Figure 5(b)**, leading the total increase in the in the freezing zone to be 15–17°C compared to 45°C in the case of A356.1 alloy. Another point to be considered from **Figure 5(c)** is the appearance of undercooling for (Al + Si) curves treated with Sr (about 2–3°C) with no effect on undercooling associated with the precipitation of α-Al phase, which is apparently controlled by addition of grain refiners [32]. Also, increasing the Ce content between 0.5 and 1.5% has no independent effect as shown in **Figure 5(c)**.

Another important aspect to be discussed is RE-grain refiner interaction. According to the Hall-Petch relationship,

$$\sigma = \mathrm{K} + 1/\mathrm{d}^{1/2} \quad (2)$$

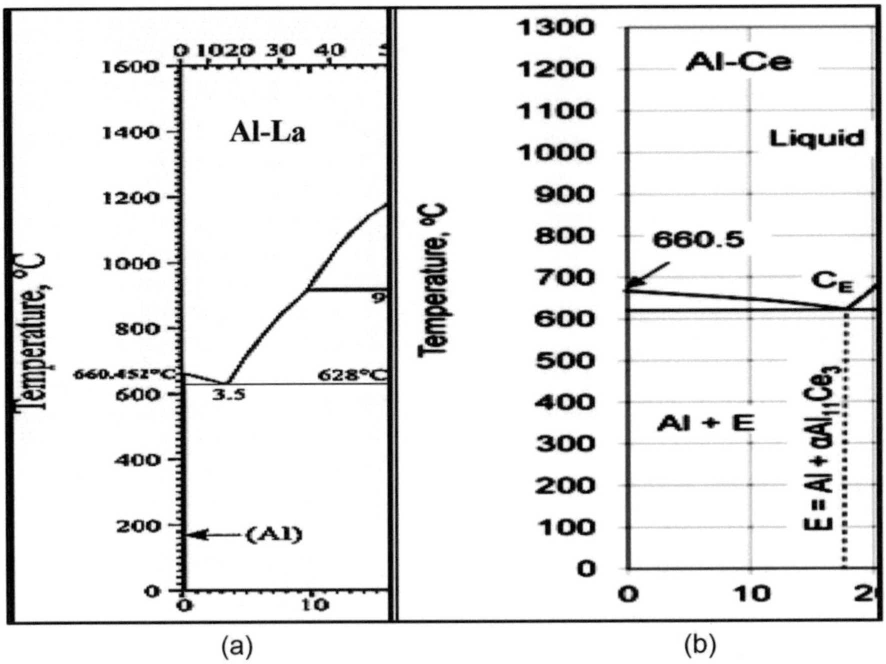

Figure 4.
Sections from (a) Al-La and (b) Al-Ce binary diagrams.

where σ = alloy strength, K is constant, and d is the average grain diameter. To emphasize this concept, the base alloy was grain refined with 0.12%Ti in the form of Al-5%Ti-1%B mater alloy added to the molten metal prior to degassing. Although the alloy originally contains 0.05%Ti in the form of TiBor, however, during remelting, the TiBor loses its effectiveness. The Ce-rich phase occurs in the form of gray compacted slugde or in star-like form (**Figure 6(a)** and **(b)**). Some authors [33–35] have defined an empirical factor (the "sludge factor"), given by:

$$\text{Sludge Factor} = \left(1\ \text{X\%Fe}\right) + \left(2\ \text{X\%Mn}\right) + \left(3\ \text{X\%Cr}\right) \quad (3)$$

When this factor becomes larger than a certain critical value, "sludge" may occur. All the elements are known as transition elements, which include Ti as well.

The X-ray distribution of Ce and Ti in one of the particles in **Figure 6(a)** are illustrated in **Figure 7(a)** and **(b)**, respectively, revealing strong intensity of Ti as confirmed by the associated EDS presented in **Figure 7(c)**, where the ratio of Ti/Ce is about 11.3/17.6 in weight % (approximately 65%). **Figure 6(c)** displays the precipitation of La-rich particles that are characterized by their shiny appearance and flat surface. As in the case of Ce, **Figure 8** reveals the X-ray distribution of La and Ti in one of the platelets in **Figure 6(c)**, where the white arrow is near the bottom of the intensity bar. The EDS pattern exhibited in **Figure 8(c)** reveals that the Ti concentration is almost 2% compared to La, which is about 40 (wt.%), that is, Ti/La is approximately 0.05%, which explains the absence of a Ti peak in the EDS spectrum.

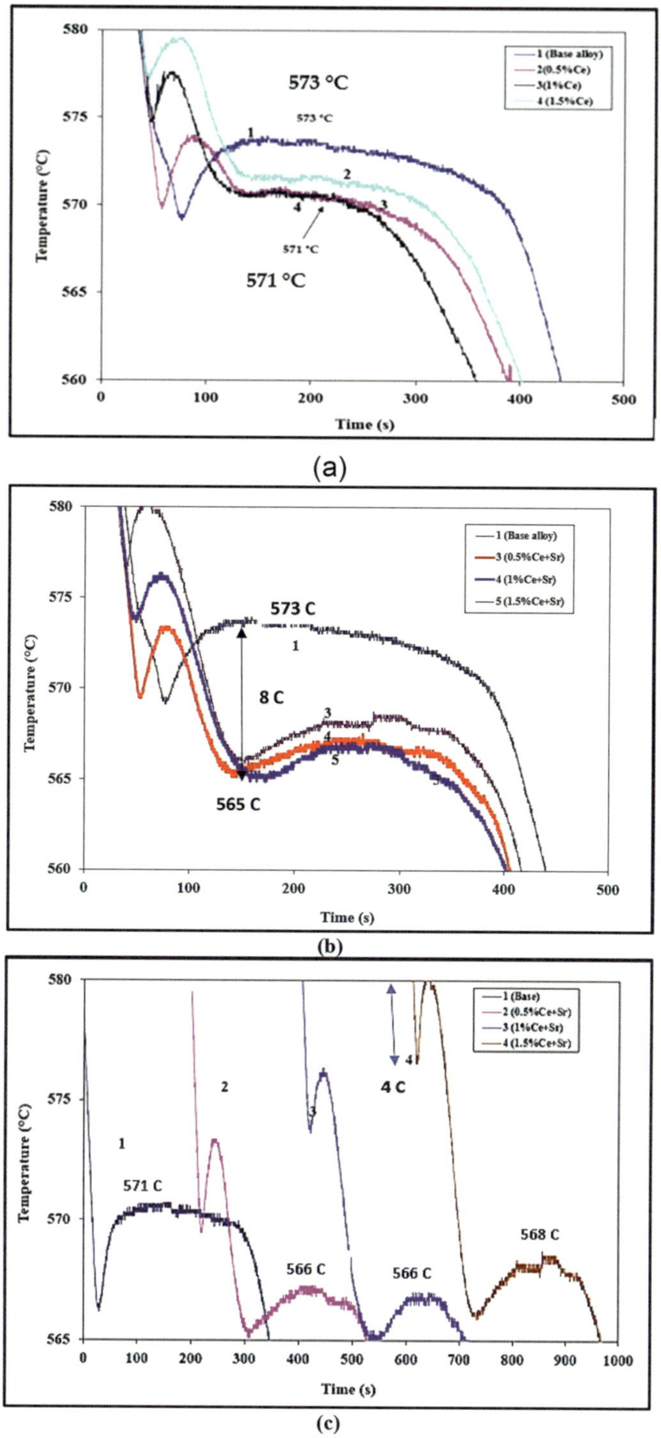

Figure 5.
(a) Effect of addition of Ce, (b) Ce + Sr on the solidification characteristics of A413.1 alloy, and (c) enlarged portions of (Al + Si) reactions in (b).

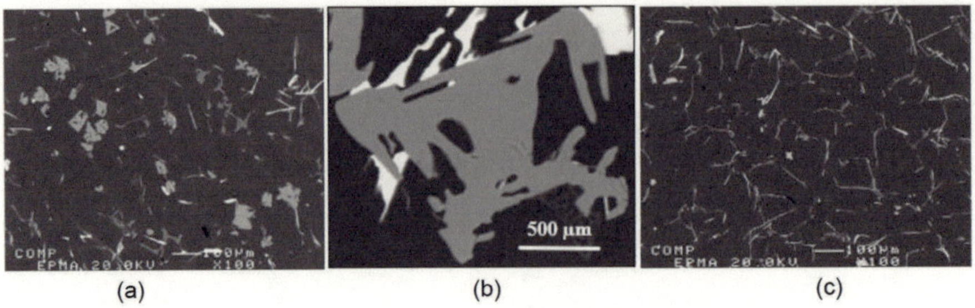

Figure 6.
Precipitation of RE intermetallics in A413.1 alloy containing: (a) 1.5%Ce; (b) high magnification image of (a); (c) 1.5%La.

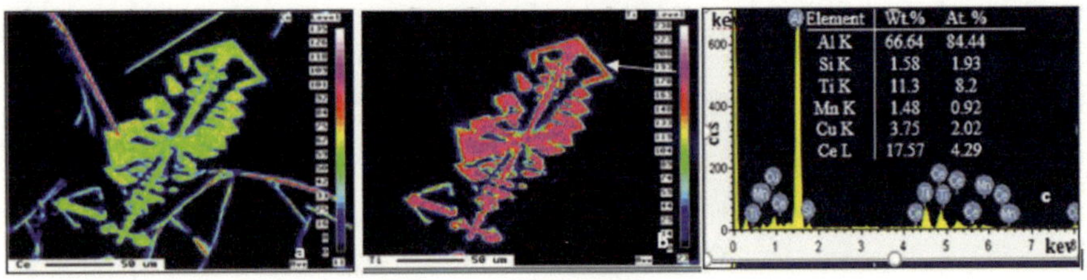

Figure 7.
X-ray images of (a) Ce, (b) Ti, (c) EDS spectrum showing the.

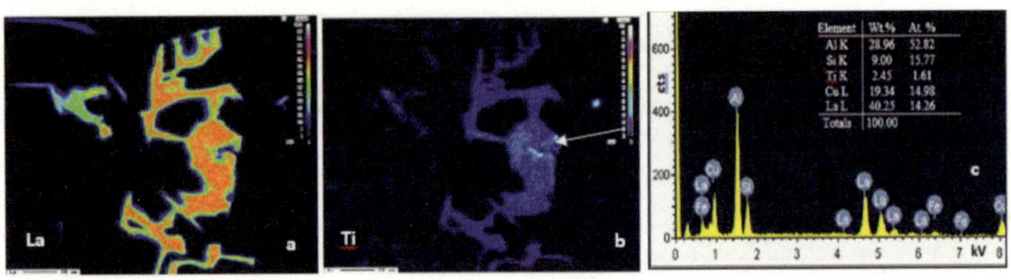

Figure 8.
X-ray images of (a) La, (b) Ti, (c) EDS spectrum showing the Ti and La concentrations.

Figure 9 shows a series of macrostructures revealing the variation in the grain size of A413.1 alloy as a function of the added Ti, La, and Ce. The average grain size of the as-received alloy was about 1850 μm (based on measuring about 100 grains using the line intercept method)—**Figure 9(a)**. Once the grain refiner was added, the average grain size was reduced to 750 μm (**Figure 9(b)**) indicating that 0.12%Ti is sufficient to refine the alloy grains. **Figure 9(c)** clearly shows the effect of Ce-Ti interaction; although the grain size was further reduced to 500 μm (**Figure 9(c)**), the combined addition of (La + Ti) resulted in a much finer grain size of about 370 μm (**Figure 9(d)**), representing 20% of the original grain size. The difference between grain size in **Figure 9(c)** and **(d)** represents the amount of Ti lost in interacting with Ce in the form of non-dissolvable intermetallics, which have a negative effect on the alloy mechanical properties, as depicted in **Figure 10** for Ce-rich intermetallics. **Figure 11** reveals a series of backscattered electron micrographs taken from the tested bars demonstrating the change in the size and distribution of the precipitated Al_2Cu phase particles, on going from fine rounded particles (**Figure 11(a)**), to short platelets (**Figure 11(b)**), to a mixture of thick platelets and spherical particles (**Figure 11(c)**)

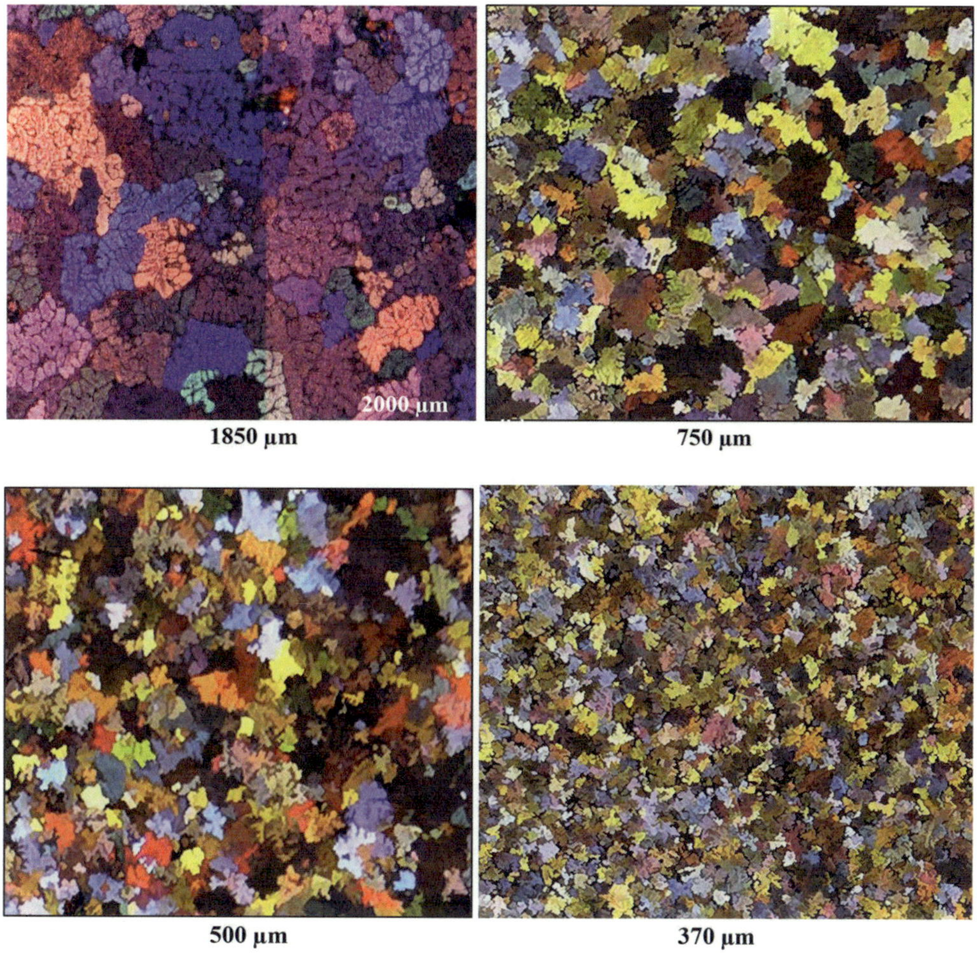

Figure 9.
Effect of grain refiner, Ce, and La addition on the alloy average grain size: (a) as-received alloy, (b) after addition of 0.12%Ti, and (c) Ti + Ce, (d) Ti + La.

indicating the beginning of incoherent precipitation. The EDS spectrum displayed in **Figure 11(d)** shows in addition to Al_2Cu particles, presence of Ce-rich intermetallics (gray phase in **Figure 11(c)**—orange arrows.

As mentioned in the experimental section, alloys for thermal analysis were not degassed due to the size of the used crucible, which reflected on porosity formation in the base alloy when RE was added. Since Ce-based phase particles precipitate in the form of star-like particles, their dendrite arms can be seen passing through the α- Al dendrite network as illustrated in **Figure 12(a)**, whereas La-rich platelets are observed lining the inner walls of the gas pores as demonstrated in **Figure 12(b)**.

3.2 Metallic variable mold

In this part, La, Ce, and La + Ce were added to the molten metal with and without Sr. **Figure 13(a)** demonstrates the occurrence of a hot zone in a small section toward the bottom of the mold (the first part to be filled with liquid metal). These zones are

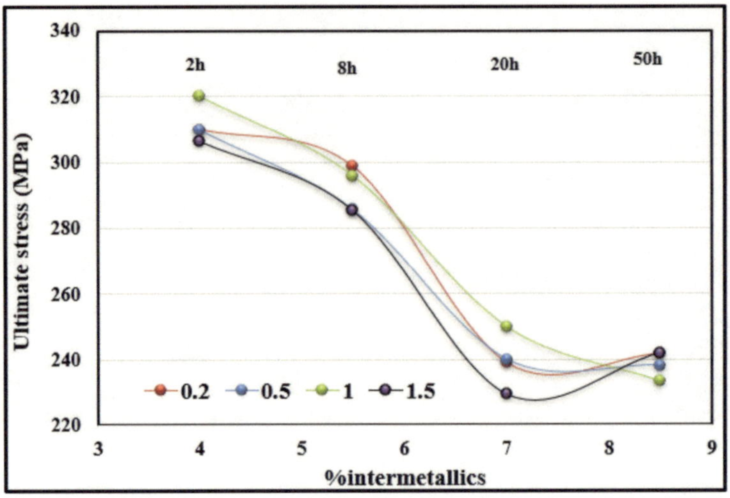

Figure 10.
Effect of aging time at 180°C and Ce concentration on the ultimate tensile strength of A413.1 alloy.

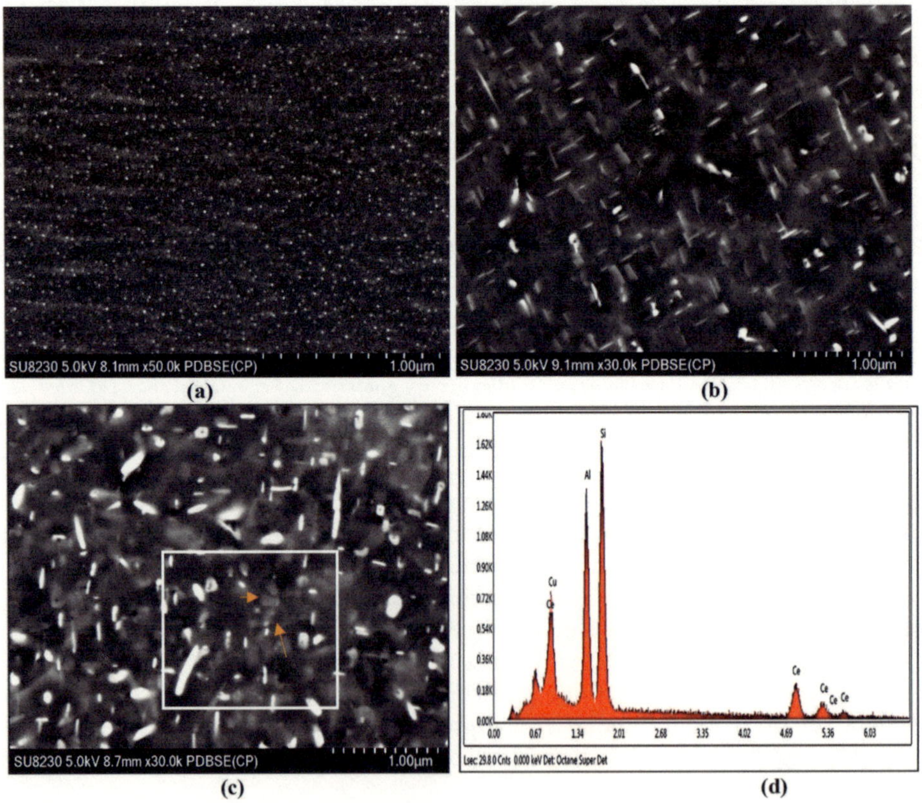

Figure 11.
Precipitation in A413.1 alloy containing 1.5%Ce and aged at 180°C: (a) 2 h, (b) 8 h, (c) 50 h, and (d) EDS spectrum corresponding to the white square in (c).

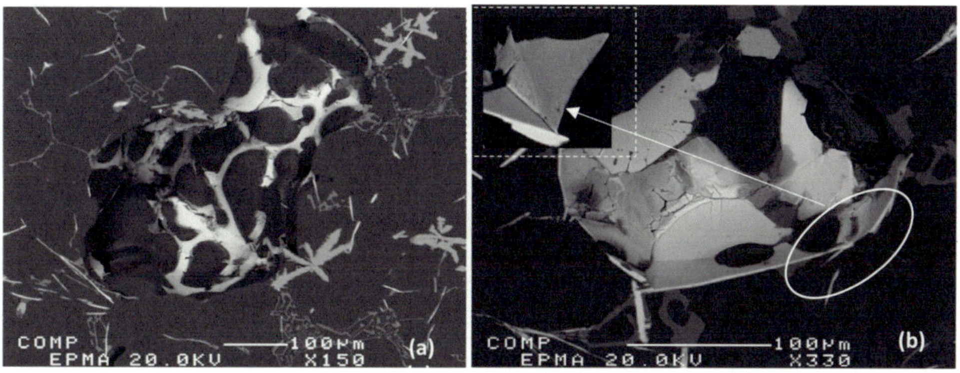

Figure 12.
Effect of RE-based intermetallics on porosity formation: (a) Ce and (b) La.

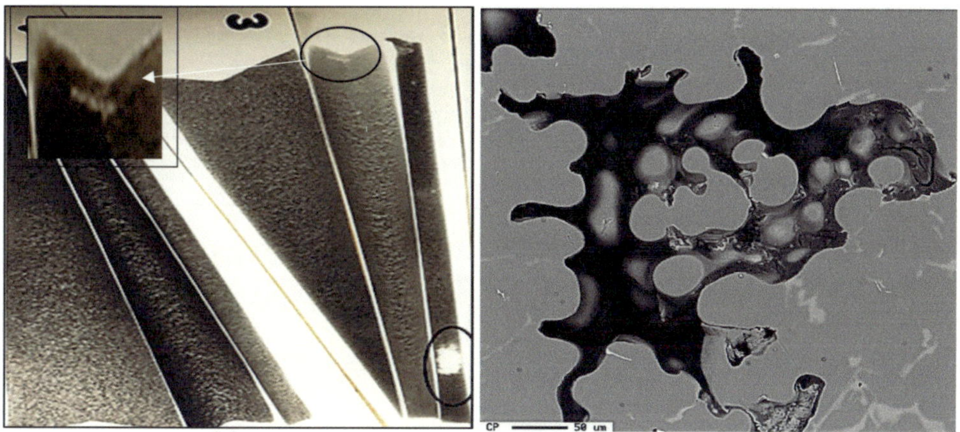

Figure 13.
(a) Radigraphs of castings made using variable metallic mold heated at 350°C. (b) Shrinkage cavities that resulted from the solidification of a hot zone.

pockets of liquid metal surrounded by solidified materials. As a result, during the solidification of these pockets, severe shrinkage cavities are formed, exemplified in **Figure 13(b)**.

Since the mold is open-face, severe shrinkage cavities would occur at the top of each casting as highlighted by the inset micrograph in **Figure 13(a)**. Also, oxide films may get trapped in the castings (SrO films), which would contribute to porosity concentration [36]. Thus, to minimize the error in evaluating porosity mainly due to RE, samples were sectioned from the middle of each casting as shown in **Figure 2(c)**. **Figure 14** depicts the effect of RE and (RE + Sr) on porosity percentage in castings at different titling angles. It should be mentioned here that the term (Ce + La) means addition of equal amounts of both elements, that is, double the amount of an individual element addition.

According to Mahmoud et al. [37–40], Ce reacts with La and other alloying elements mainly Al, Fe, Sr, and Cu. Although the added amount of RE is double that

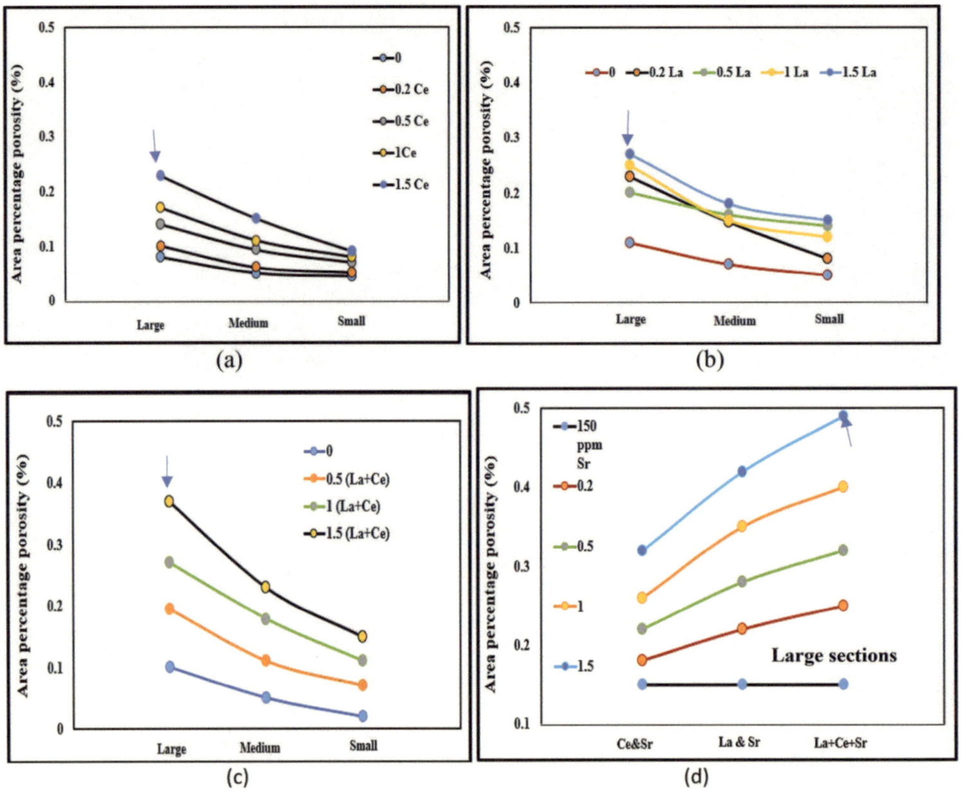

Figure 14.
Effect of RE on porosity percentage: (a) Ce, (b) La, (c) Ce + La, and (d) RE + Sr—arrows point to a large section (maximum % of porosity).

of a single element addition, the resulting volume fraction of intermetallics is higher than that obtained from individual elements but not necessarily double the volume fraction shown in **Figure 10**. **Figure 14(a)** and **(b)** display the variation in % of total porosity as a function of casting section and added RE. Both Ce and La addition in the same quantity produces the same pattern except for the addition of La, where it may result in a slightly higher value (0.28%) than that reported for Ce (0.22%) due to the difference in the morphology of the precipitated intermetallics, that is, platelets vs. star-like. Doubling the concentration—(La + Ce)—**Figure 14(c)** increased the % of porosity to about 0.42% (large sections). As shown in **Figure 14(d)**, treating the alloy with 150 ppm Sr led to a noticeable increase in % porosity, up to 0.5%, representing a mixture of porosity caused by RE and Sr.

In order to gain a better understanding of the data presented in **Figure 14**, a series of backscattered electron taken from large section samples is shown in **Figure 15** demonstrating the progressive increase in % porosity with the added elements. **Figure 15(a)** represents the base alloy revealing the absence of porosity, whereas **Figure 15(b)** of A413.1 alloy containing 1.5%La exhibits the appearance of scattered shrinkage cavities shown in areas marked A, B, and C. As presented in **Figure 5**, the maximum melting point of the present alloy containing 1.5%La is in the range of 575–580°C, which is much lower than the eutectic reactions depicted in **Figure 4**. Thus, it is reasonable to assume that La- and Ce-rich intermetallics may be precipitated prior

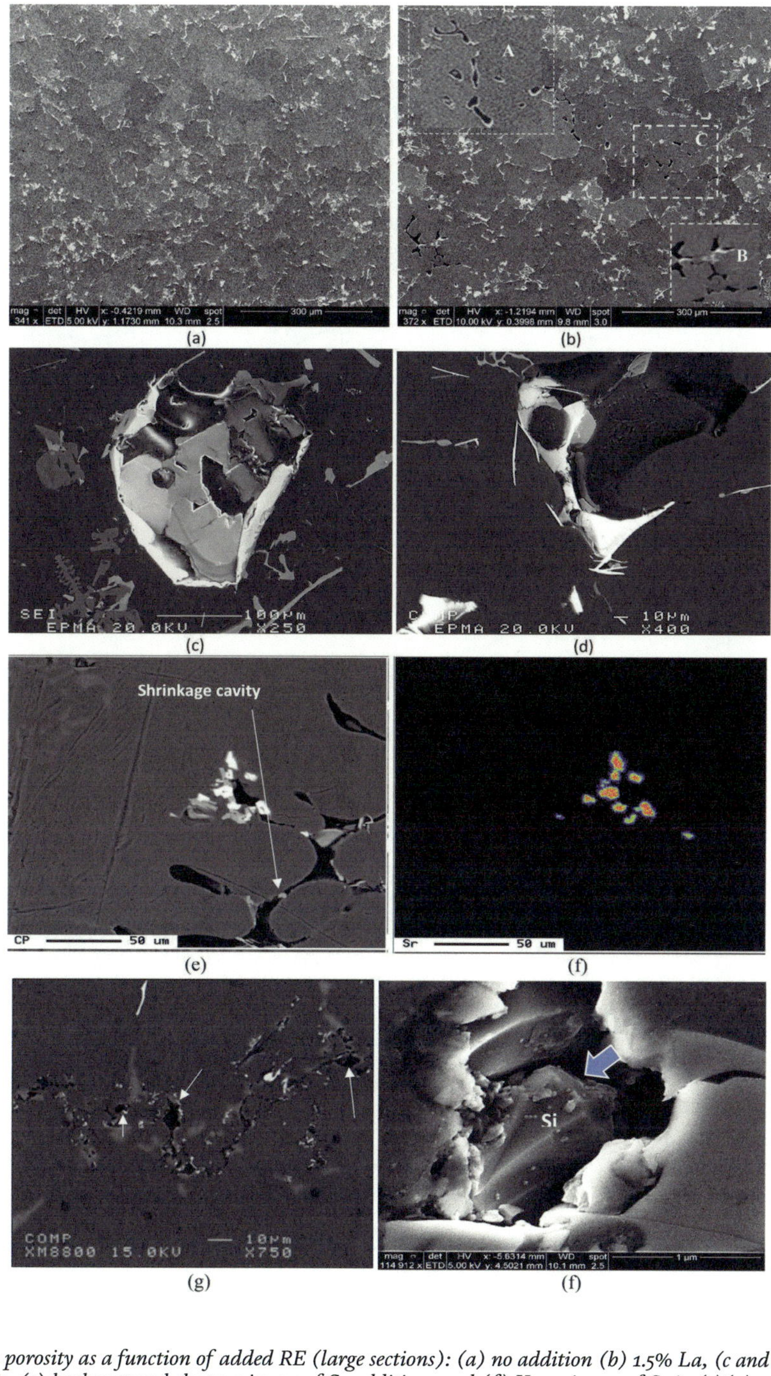

Figure 15.
Variation in porosity as a function of added RE (large sections): (a) no addition (b) 1.5% La, (c and d) 1.5% La + 1.5% Ce, (e) backscattered electron image of Sr addition, and (f) X-ray image of Sr in (e)(g) porosity associated with a long chain of oxide films-white arrow (f) morphology of an eutectic Si particle in a Sr-modified alloy—blue arrow.

to the formation of the α-Al network, leading to blocking the flow of liquid metal and hence explaining the large pore shown in **Figures 15(c)** and **(d)**. **Figure 15(e)** and **(f)** are good examples of precipitation of Sr-rich components in the aluminum matrix in

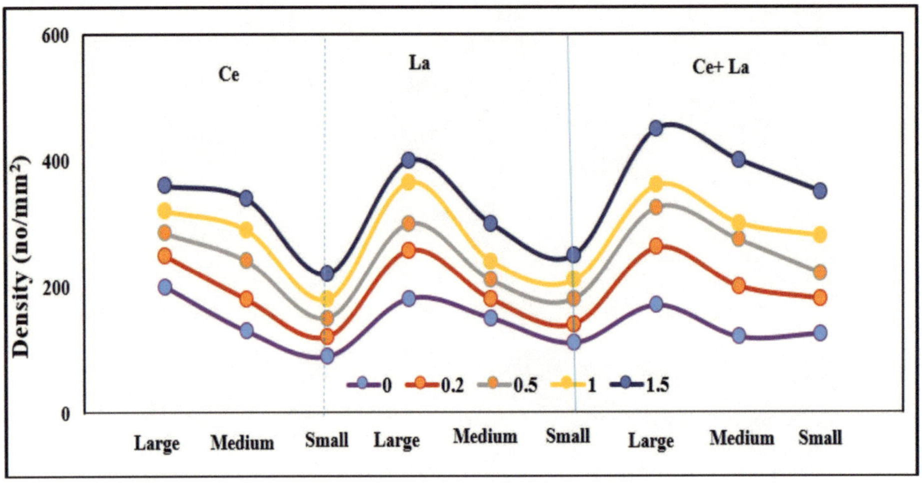

Figure 16.
Variation of density of porosity as a function of casting angle and added RE.

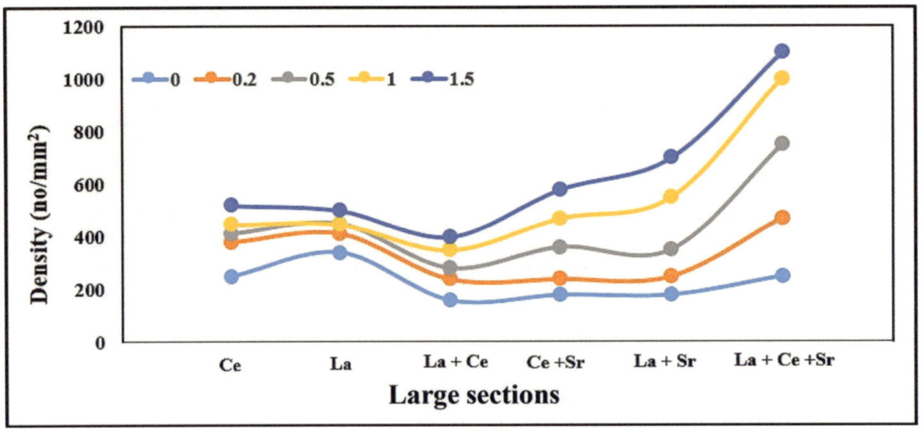

Figure 17.
Variation of density of porosity as a function of added RE and Sr.

the vicinity of large shrinkage cavities. Since there was no filtering, oxides can easily slip into the melt and create fine pores along their lengths as displayed in **Figure 15(g)**. An example of a modified Si particle in Sr treated alloy is given in **Figure 15(f)**.

Another parameter to be considered is the density of porosity caused by addition of the above elements as illustrated in **Figure 16**. The curve is divided into three blocks separated by thin broken lines. In each case, the density systematically increases as the mold angle increases, and the total amount of the added RE (no modification was used) reaching a maximum of 420 pores/mm^2 (1.5%La + 1.5%Ce, large angle). With the addition of 150 ppm Sr into the molten metal, a significant steady increase took place with the increase in the added materials reaching almost 1100 pores/mm^2—**Figure 17**, which is almost three times what was observed in **Figure 16**.

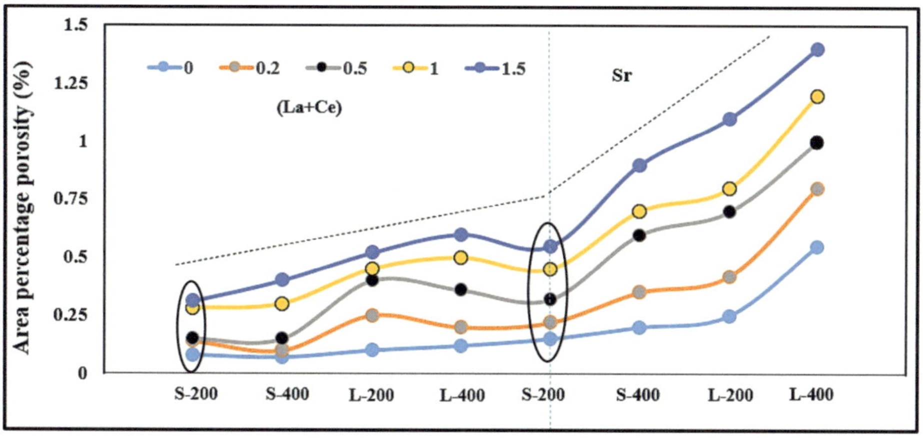

Figure 18.
Variation of area percentage porosity as a function of mold temperature, sample section, and RE.

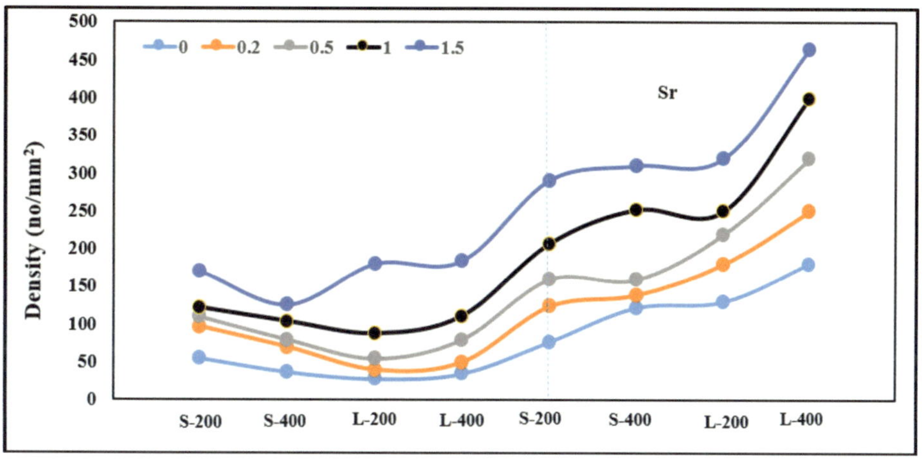

Figure 19.
Variation of pore density as a function of mold temperature, sample section, and RE.

3.3 Step-like metallic mold

As shown in **Figure 2(d)**, the step-like mold is made of thick walls (10 mm) with a pore cavity almost half the weight of the total casting to minimize occurrence of shrinkage cavities within the casting itself. Also, 1 mm from the top of sectioned samples was removed to avoid any possibility of interfering with casting shrinkage cavities with those resulting from the addition of RE (La + Ce). Both **Figures 18** and **19** reveal that aside from the increase of the parameter with the amount of introduced materials, a marked observation is that the slope of the curve is divided into two distinct angles when the melt was modified with Sr (150–180 ppm), in large sections, similar to those exhibited in **Figure 17**. The black circle in **Figure 18** shows a relatively high % porosity in small sections heated at 200°C, which may in part be due to reduction in the fluidity of molten metal in the non-modified alloys.

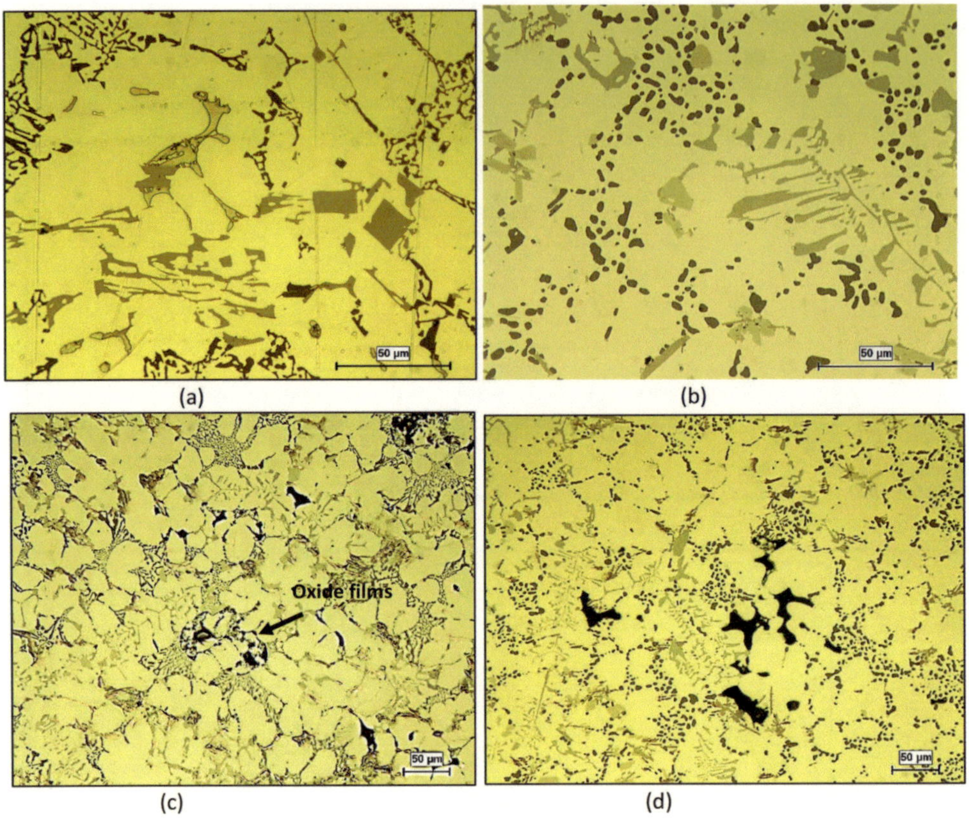

Figure 20.
Optical micrographs of RE and Porosity in large sections at 400°C: (a) precipitation of RE intermetallics in non-modified alloy, (b) precipitation of RE intermetallics in Sr-modified alloy, (c) porosity formation in non-modified alloy, and (d) porosity formation in Sr-modified alloy.

Figure 20 exhibits optical micrographs taken from large sections (400°C). **Figure 20(a)** and **(b)** show the precipitation of RE in non-modified and Sr-modified alloys, respectively-light gray phases (1.5%La +1.5%Ce). Since the molten metal was degassed prior pouring into the mold, most of the porosity is caused due to shrinkage. The pore size is small due to the use of a large filling cavity. Since no filtration was applied, some oxide films were able to pass through as shown in **Figure 20(c)**.

4. Conclusion

Based on the present results on the effect of additives and casting process on porosity formation in A413.1, the following conclusions may be drawn:

1. The addition of 1.5% rare earth metals (RE) largely increases the melting point of A413.1 alloy by 5-7 degrees due to the presence of new alloying elements, which is independent of Sr-modification.

2. RE reduces the eutectic temperature by about 3-4 degrees in non-modified alloys and 9°C in Sr-modified ones. There is no difference between 0.5% and 1.5% RE addition.

3. Increasing %RE has a marked deterioration on the alloy strength due to large volume fraction of insoluble intermetallics.

4. Due to the high affinity of Ce to react with Ti, the addition of 1.5%Ce would poison the grain refining. On the other hand, La has no reaction with Ti, and hence addition of 0.12%Ti + 1.5%La would lead to about 80% reduction in the alloy grain size.

5. Most of the observed porosities are caused by shrinkage, especially in Sr-treated alloys, regardless of the mold type.

6. The observation that some of the round pores are lined by layers of RE intermetallics may suggest that these intermetallics were precipitated in the liquid state blocking the flow of the liquid metals (maximum alloy melting point is about 580°C).

Author details

Ehab Samuel[1], Agnes M. Samuel[1], Victor Songmene[2] and Fawzy H. Samuel[1*]

1 Département des Sciences Appliquées, Université du Québec à Chicoutimi, Chicoutimi, QC, Canada

2 Department of Mechanical Engineering, École de Technologie Supérieure (ÉTS), Montréal, QC, Canada

*Address all correspondence to: fhsamuel@uqac.ca

References

[1] Humphries, Rare Earth Elements: The Global Supply Chain, Congressional Research Service. 2013

[2] Long KR, Van Gosen BS, Foley NK, Cordier D. The Geology of Rare Earth Elements, The Principal Rare Earth Elements Deposits of the United States, USGS Scientific. 2010

[3] King HM. REE - Rare Earth Elements and their Uses, Geology.com, Geoscience news and information. 2023

[4] Cordier DJ. Rare Earths. Mineral Commodity Summaries: United States Geological Survey; 2023

[5] Balaram. Rare earth elements: A review of applications, occurrence, exploration, analysis, recycling, and environmental impact, Geoscience Frontiers. Jul 2019;**10**(4):1285-1303

[6] Gursoy O, Timelli G. Lanthanides: A focused review of eutectic modification in hypoeutectic Al–Si alloys. Journal of Materials Research and Technology. 2020;**9**:8652-8666

[7] Jha K, Kumari A, Panda R, Kumar JR, Yoo K, Lee JY. Review on hydrometallurgical recovery of rare earth metals. Hydrometallurgy. 2016;**165**:2-26

[8] Tunsu C, Ekberg C, Foreman M, Retegan T. Studies on the solvent extraction of rare earth metals from fluorescent lamp waste using cyanex 923. Solvent Extraction and Ion Exchange. 2014;**32**:650-668. DOI: 10.1080/07366299.2014.925297

[9] Anastopoulos P, Bhatnagar A, Lima EC. Adsorption of rare earth metals: A review of recent literature. Journal of Molecular Liquids. 2016;**221**:954-962

[10] Miur T, Wada A. Precise purity analysis of high-purity lanthanum oxide by gravimetric analysis assisted with trace elemental analysis by inductively coupled plasma mass spectrometry. Frontiers in Chemistry. 2022;**10**:888636

[11] Abbasalizadeh A, Malfliet A, Seetharaman S, Sietsma J, Yang Y. Electrochemical extraction of rare earth metals in molten fluorides: Conversion of rare earth oxides into rare earth fluorides using fluoride additives. Journal of Sustainable Metallurgy. 2017;**3**:627-637. DOI: 10.1007/s40831-017-0120-x

[12] Zhou B, Li Z, Chen C. Global potential of rare earth resources and rare earth demand from clean technologies. Minerals. 2017;7(11):203. DOI: 10.3390/min7110203

[13] Zhu M, Jian ZY, Yang GC, Zhou YH. Effects of T6 heat treatment on the microstructure, tensile properties, and fracture behavior of the modified A356 alloys. Materials & Design. 2012;**36**:243-249

[14] Mousavi GS, Emamy M, Rassizadehghani J. The effect of mischmetal and heat treatment on the microstructure and tensile properties of A357 Al-Si casting alloy. Materials Science and Engineering a-Structural Materials Properties Microstructure and Processing. 2012;**556**:573-581

[15] Dang B, Jian ZY, Xu JF. Effects of rare-earth element addition and heat treatment on the microstructures and mechanical properties of Al-25% Si alloy. International Journal of Materials Research. 2017;**108**:269-274

[16] Zhang X, Wang ZH, Zhou ZH, Xu JM. Influence of rare earth (Ce and

La) addition on the performance of Al-3.0 wt%Mg alloy. Journal of Wuhan University of Technology-Materials Science Edition. 2017;**32**:611-618

[17] Xiao DH, Wang JN, Ding DY, Yang HL. Effect of rare earth Ce addition on the microstructure and mechanical properties of an Al–Cu–Mg–Ag alloy. Journal of Alloys and Compounds. 2003;**352**:84-88

[18] Xiao DH, Wang JN, Ding DY. Effect of minor cerium additions on microstructure and mechanical properties of cast Al – Cu – Mg – Ag alloy. Materials Science and Technology. 2004;**20**:1237-1240

[19] Tsai YC, Lee SL, Lin CK. Effect of trace Ce addition on the microstructures and mechanical properties of A356 (AL-7SI-0.35 Mg) aluminum alloys. Journal of the Chinese Institute of Engineers. 2011;**34**:609-616

[20] Voncina M, Kores S, Mrvar P, Medved J. Effect of Ce on solidification and mechanical properties of A360 alloy. Journal of Alloys and Compounds. 2011;**509**:7349-7355

[21] Voncina M, Mrvar P, Petric M, Medved J. Microstructure and grain refining performance of Ce on A380 alloy. Journal of Mining and Metallurgy, Section B: Metallurgy. 2012;**48**:265-272

[22] Li QL, Xia TD, Lan YF, Zhao WJ, Fan L, Li PF. Effect of rare earth cerium addition on the microstructure and tensile properties of hypereutectic Al-20%Si alloy. Journal of Alloys and Compounds. 2013;**562**:25-32

[23] Ye LY, Gu G, Liu J, Jiang HC, Zhang XM. Influence of Ce addition on impact properties and microstructures of 2519A aluminum alloy. Materials Science and Engineering: A. 2013;**582**:84-90

[24] Yii SLJ, Anas NM, Ramdziah MN, Anasyida AS. Microstructural and mechanical properties of Al-20%Si containing cerium. Procedia Chemistry. 2016;**19**:304-310

[25] Ouyang ZY, Mao XM, Hong M. Multiplex modification with rare earth elements and P for hypereutectic Al-Si alloys. Journal of Shanghai University (English Edition). 2007;**11**:400-402

[26] Tsai YC, Chou CY, Lee SL, Lin CK, Lin JC, Lim SW. Effect of trace La addition on the microstructures and mechanical properties of A356 (Al–7Si–0.35 Mg) aluminum alloys. Journal of Alloys and Compounds. 2009;**487**:157-162

[27] Hosseinifar M, Malakhov DV. The sequence of intermetallics formation during the solidification of an Al-Mg-Si alloy containing La. Metallurgical and Materials Transactions A. 2010;**42**:825-833

[28] Mahmoud MG, Samuel AM, Doty HW, Valtierra S, Samuel FH. Effect of rare earth metals, Sr, and Ti addition on the microstructural characterization of A413.1 alloy. Advances in Materials Science and Engineering. 2017;**2017**:12. DOI: 10.1155/2017/4712946

[29] Alkahtani SA, Elgallad EM, Tash MM, Samuel AM, Samuel FH. Effect of rare earth metals on the microstructure of Al-Si based alloys. Materials. 2016;**9**(1):45

[30] Elgallad EM, Doty HW, Alkahtani SA, Samuel FH. Effects of La and Ce addition on the modification of Al-Si based alloys. Advances in Materials Science and Engineering. 2016;**2016**:13

[31] Cacciamani G, Ferro R. Thermodynamic modeling of some aluminum-rare earth binary systems:

Al-La, Al-Ce and Al-Nd. Calphad. 2001;**25**(4):583-597

[32] Nabawy AM, Samuel AM, Alkahtani SA, Abuhasel KA, Samuel FH. Role of cerium, lanthanum, and strontium additions in an Al–Si–Mg (A356) alloy. International Journal of Materials Research. 2016;**107**(5):446-458

[33] Ferraro S, Fabrizi A, Timelli G. Evolution of sludge particles in secondary die-cast aluminum alloys as function of Fe, Mn and Cr contents. Materials Chemistry and Physics. 2015;**153**(1):168-179

[34] Dong X, Ji S. Si poisoning and promotion on the microstructure and mechanical properties of Al–Si–Mg cast alloys. Journal of Materials Science. 2018;**53**:7778-7792

[35] Weiler JP. A review of magnesium die-castings for closure applications. Journal of Magnesium and Alloys. 2019;**7**(2):297-304. DOI: 10.1016/j.jma.2019.02.005

[36] Samuel AM, Ammar HR, Zedan Y, Doty HW, Samuel FH. Effect of Ca–Sr–Mg and Bi–Sr–Mg interactions on the microstructural characterization and tensile properties of B319 alloy. International Journal of Metalcasting. 2022;**16**(4):1940-1959

[37] Mahmoud MG, Zedan Y, Samuel AM, Songmene V, Samuel FH. The use of rare earth metals in Al–Si–Cu casting alloys. International Journal of Metalcasting. 2022;**16**(2):535-552

[38] Mahmoud MG, Zedan Y, Samuel AM, Songmene V, Doty HW, Samuel FH. Applications of Rare Earth Metals in Al-Si Cast Alloys, in Advances in High-Entropy Alloys - Materials Research, Exotic Properties and Applications. IntechOpen; 2021

[39] Mahmoud MG, Samuel AM, Doty HW, Samuel FH. Formation of Rare Earth Intermetallics in Al–Cu Cast Alloys, in Light Metals Symposium held at the 149th Annual Meeting and Exhibition. San Diego; United States: TMS 2020; 2020. pp. 241-246

[40] Makhlouf MM, Apelian D. Casting Characteristics of Aluminum Die Casting Alloys. Vol. No. DOE/ID/13716. Worcester, Massachusetts, USA: Worcester Polytechnic Institute; 2002

Chapter 7

The Effect of Chemical Composition on EN AW 6XXX Series Aluminum Alloys

Emrah Fahri Ozdogru

Abstract

6XXX Series Aluminum alloy is an Aluminum–Magnesium–Silicon family (6XXX series) alloy which is the most common and widely used in many sectors such as automobile, aircraft, marine and construction, owing to its low cost, high strength to weight ratio, good formability, weldability, excellent corrosion resistance, and higher thermal conductivity. The principal alloying elements in 6XXX series are Silicon (Si) and Magnesium (Mg) and they are controlling the mechanical properties of the alloy with the precipitation on heat treatment process. Considering all elements used in 6XXX alloys, the main objective for alloy development is to get high productivity with superior mechanical properties which means to create alloy chemistry for having easy flow during extrusion, less quench sensitivity, and high mechanical properties with short aging cycle after extrusion process. In this chapter, Alloy developments for aluminum alloys which is mainly used for automotive (6063 and 6082 alloy will be the main target) will be discussed and industrial practice results will be given. Many Optical microscope and SEM (Scanning Electron Microscope) pictures will be shared coming from the lab and industrial scale works. The characterization method to see the effect of alloy elements will be also one of the focuses on this chapter.

Keywords: aluminum alloys, extrusion, 6063 and 6082 aluminum alloys, microstructure, chemical composition

1. Introduction

Aluminum is third most abundant elements (the most abundant as a metal) found in the earth crust as bauxite. It has a face centered cubic crystal structure and it is a silvery-white, soft, ductile, and non-magnetic metal. The physical and mechanical of aluminum makes it the most broadly used metal after steel on the market. Aluminum is one of the lightest engineering metals, having a strength to weight ratio higher to steel and similar approaches for conductor cables against the copper.

Aluminum in its pure forms relatively soft and for this reason, the aluminum is alloyed with a range of different elements to achieve required mechanical and physical properties. The most important property of aluminum is its light weight with having density 2.7 g/cm^3, which is about one-third of steel.

Aluminum alloys have excellent casting application with its low melting point, high degree of fluidity in molten stage. Structural materials require not only high strength to weight ratio, besides reasonable cost, but also high fatigue resistance. Aluminum alloys have some other attributes like good electrical and thermal conductivities, corrosion resistance, good workability, ductility, and strength which make aluminum alloy widely usable and useful materials. Because of its good formability and workability, it can be easily forged and extruded to any shape and size. Aluminum and its alloys when exposed to atmosphere, combines with oxygen to form a protective oxide coating which blocks further oxidation and protect the surface especially on the DC casting application. The formation of protective oxide coating makes it highly corrosion resistant. If the protective layer of aluminum is scratched, it will instantly reseal itself [1, 2].

Among the aluminum alloys, 6XXX series aluminum alloy which is the Al-Mg-Si family are the most common group and seen on the engineering applications such as construction, automobile, aircraft, marine and railway due to its high mechanical strength, corrosion resistance. 6XXX series commercial aluminum alloys do not promote the mechanical properties like either 2XXX or 7XXX series alloys, but the range is one of the most versatile heat-treatable aluminum and offer good corrosion resistance and formability with medium strength. The ability to improve their mechanical properties is very clear in 6XXX series. The reason behind the use of aluminum alloys in so many different purposes is the element diversity in the chemical composition and the ability to adjust the heat treatments as desired.

The requirements for producing high quality 6XXX series aluminum alloys are inevitably associated with the precision alloy design, innovative production technology and use versatile heat-treatment practices and enhancement of mechanical properties [3–5].

The principal alloying elements in 6XXX series are Silicon (Si) and Magnesium (Mg) and they are controlling the mechanical properties of the alloy with the precipitation on heat treatment process. The 6XXX series aluminum alloys are the best alloy for the extrusion process and one the most common alloys for extrusion application due to wide range profile production and focus for many applications such as construction, home appliances such as kitchen and furniture, electricity purposes, industrial application and automotive, rail and aerospace.

The focus of this review paper is to put together the latest knowledge available from various sources on alloy designation, industrial processing, development of properties and potential use of AA6XXX wrought alloys.

2. Chemistry and Its effect on phase formation in 6XXX aluminum alloys

The pure aluminum is very soft and easy to form. However, it is not possible to use in any engineering application. Therefore, aluminum is alloyed in order to increase strength. On 6XXX series aluminum alloys, the Mg and Si are the major alloying elements. The main strengthening mechanisms are precipitation or age hardening, work hardening and grain boundary strengthening. The purpose of adding Mg and Si is to from MgxSiy compound (stochiometric relationship will be detailed on further pages) inside the grain. The other popular alloying elements are Mn, Cr, Cu, Zr, Sc and Ti (mostly as a grain refiner). Fe is a trace element and always coming from the mineral and production of the pure aluminum. The final purpose of the alloy design is to establish a perfect microstructure to achieve best strength, ductility and toughness

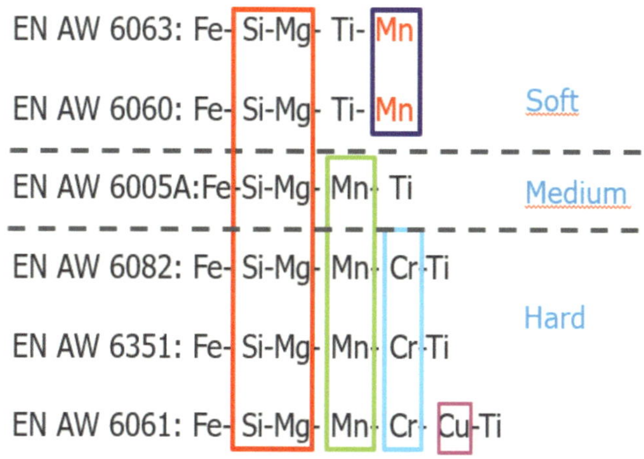

Figure 1.
Specific alloying elements in some common 6XXX series aluminum alloys. (Image copyright © belongs to TRI METALURJI A.S.).

along with other properties like corrosion and fatigue. **Figure 1** shows the summary of alloying elements in some common 6XXX series aluminum alloys.

By increasing the Mg and Si content on the alloy system, the mechanical strength improves and but also, the other applications such as process response, aging, and surface treatment will be affected. The well-known 6XXX series alloy grade according to Mg/Si ratio is given in **Figure 2**.

Mg and Si are major alloying elements and easily dissolute on the aluminum solid solution during solution heat treatment process [6]. Magnesium (Mg) and silicon (Si) which combine to form the Mg_2Si precipitates. These precipitates occur in several forms which may be divided into the following three categories.

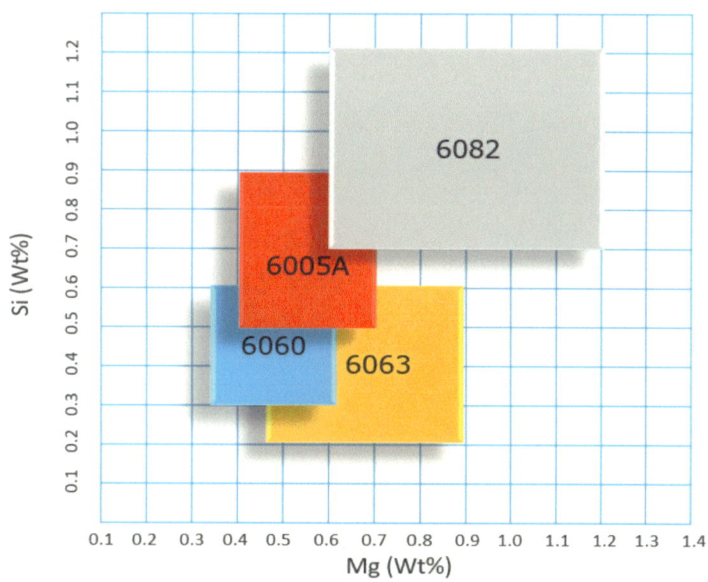

Figure 2.
Mg/Si ratio vs. alloy grades (Image copyright © belongs to TRI METALURJI A.S.).

$$SSSS \rightarrow \text{solute clusters} \rightarrow \text{GP-zones} \rightarrow \beta'' \rightarrow \beta' \rightarrow \beta\ (Mg_2Si).$$

In details;

- β" (beta double prime) Mg_2Si, the smallest type of Mg_2Si precipitate that is rod-shaped and contributes most to mechanical properties when densely dispersed.
- β' (beta prime) Mg_2Si, a larger version of rod-shaped precipitate that grows from the β" category. The β' precipitates have a negligible contribution to mechanical properties.
- β (beta) Mg_2Si, the largest Mg_2Si precipitate that is cube-like in shape and due to its size Its not affecting to mechanical properties. **Figure 3** shows typical view of Mg_xSi_y precipitation in 6082 aluminum alloys.

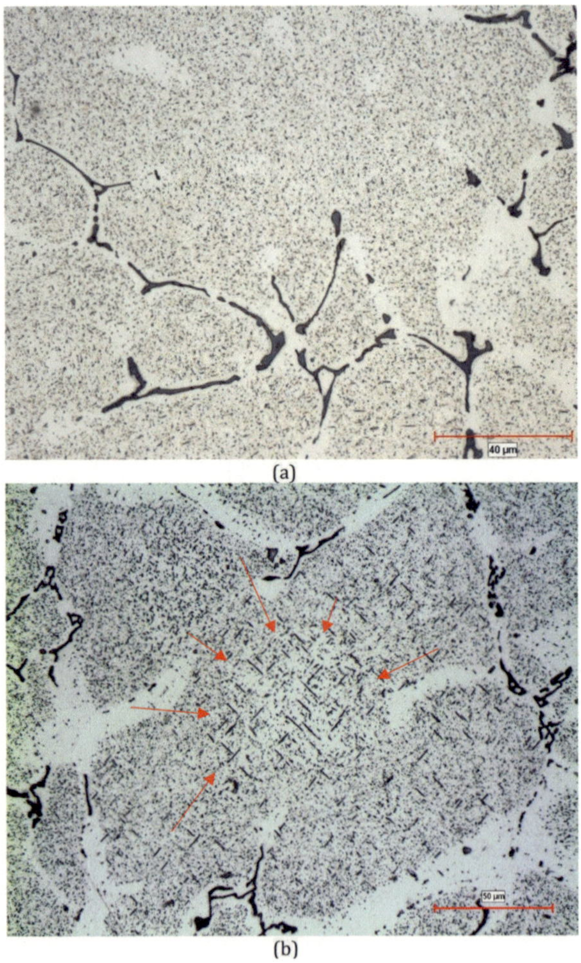

Figure 3.
Mg_2Si intermetallic formation in 6082 aluminum alloys a) fine and well distributed and b) coarse and heterogenous distributed (the pictures were taken from the billet slice after quenching process) (a) Fine Mg2Si (β" intermetallics-beta double prime-) (b) Coarse Mg2Si (β -beta-), (Image copyright © belongs to TRI METALURJI A.S.).

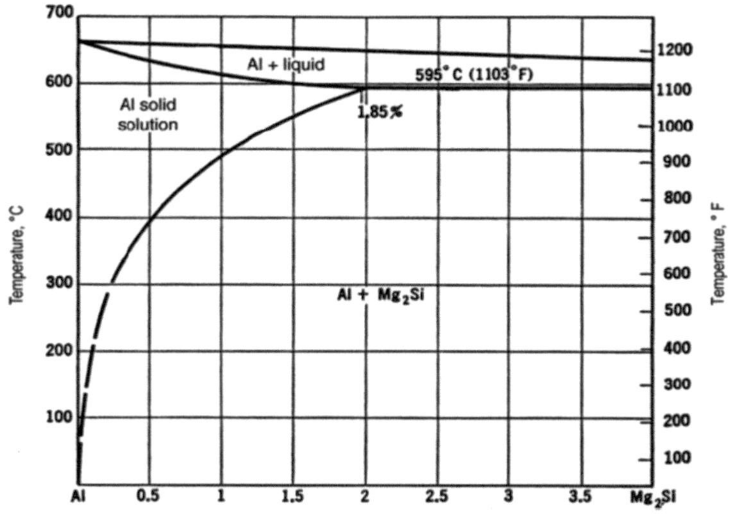

Figure 4.
Al-Mg_2Si pseudo binary phase diagram [7].

The proper ratio for Mg_2Si is Mg/Si = 1.73, but this is impossible to achieve with ordinary operating tolerances; thus, most alloys have either magnesium or silicon excess. Magnesium excess leads to better corrosion resistance but lower strength and formability; silicon excess produces higher strength without loss of formability and weldability, but there is some tendency to intergranular corrosion.

The formation of Mg_2Si precipitates give rise to the simple eutectic system with aluminum. At elevated temperature, the solute element dissolves in the solid solution but because of the decrease in solubility at lower temperature forms age-hardenable Mg_2Si precipitates. **Figure 4** shows the Al-Mg_2Si phase diagrams. According to the phase diagram, with increasing amount of Mg and Silicon, peritectic point which means Mg_2Si precipitation formation starts to decrease the eutectic reaction starts earlier than the expectation.

On the practical application such as casting, homogenization and extrusion process, the temperature control is becoming vital parameters to achieve mechanical strength and better surface properties of the aluminum material.

In other words, looking at the alloy simulation program such as JMAT-PRO®, It's also inevitable seen that the evaluation of Mg_2Si formation itself is not enough, since the formation of excess silicon or excess magnesium condition. By increasing the Mg and Si content on the alloy system, the peritectic point shifts to the aluminum solid solution side and peritectic points (temperature) also reduced with some point. **Figure 5** shows two different Mg, Si content aluminum alloy peritectic point evaluation (on the phase diagram, The Iron content was not taking into account).

Figure 5 shows clearly that why EN AW 6082 aluminum alloy is more sensitive for production and why it should be extruded in low speed. The low eutectic point phases restrict the extrusion speed and not allowed to go up higher production speed. Precipitation of the alloying elements as coherent GP zones and coherent β'' and hexagonal coherent β' (Mg_2Si), during aging treatment, provides strengthening due to the presence of coherency strain field around the precipitates, which interacts with the moving dislocations. Formation of nonequilibrium coherent hexagonal precipitation gives the highest precipitation strengthening in AA6XXX alloys. The typical

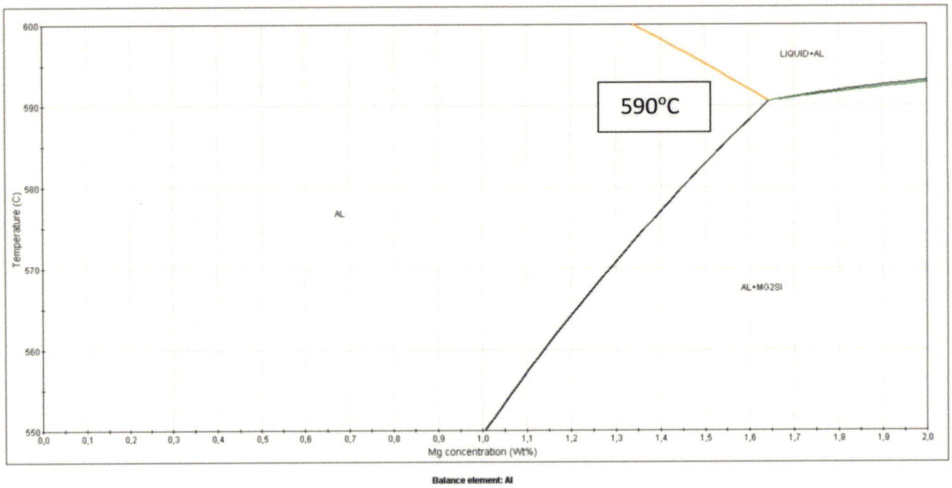

(a)

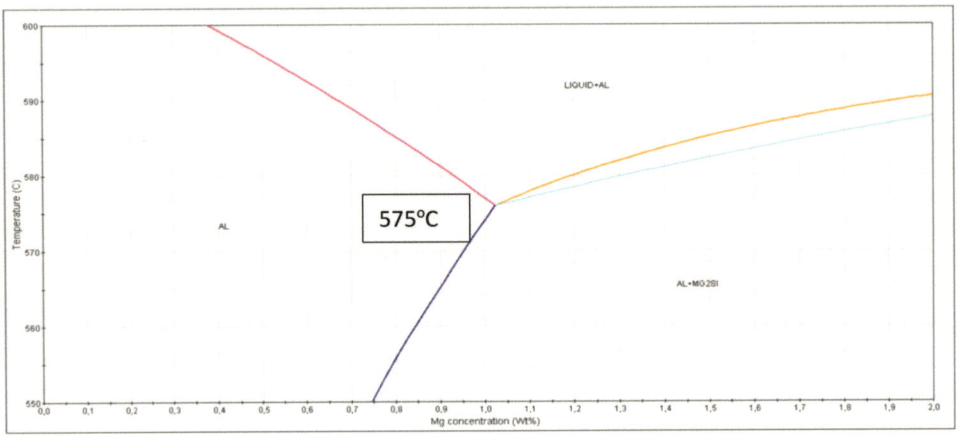

(b)

Figure 5.
Two main 6XXX series aluminum alloys peritectic reaction (a) Mg: 0.35% and Si: 0.50% (typical 6060 Al Alloys) (b) Mg: 0.7% and Si: 09% (typical 6082 Al alloys).

phase transformation analysis by using DSC for standard EN AW 6082 aluminum alloys during again is given in **Figure 6**.

Iron (Fe) is also present in the alloys and combines with silicon (Si) and aluminum (Al) to form AlFeSi dispersoids. These intermetallic do not contribute to the strength of the alloy but, if they are not correctly processed, they will have a detrimental effect on the extrudability of the alloy. Accurate control of Fe contents in 6000 series alloys is important for surface finishing applications. Different levels of Fe will cause variations in glossiness response during anodizing. Fe is also known to reduce conductivity. As Its well known, the homogenization process is applied to DC casting billet in order to provide phase transformation for Fe based dispersoid. During homogenization, the needle shape Kβ-AlFeSi dispersoid transform into cubic shape. **Figure 7** shows before and after homogenization microstructure of EN AW 6063 aluminum alloys.

DC casting is the most common and well-known method for 6XXX series. On the metallurgy of 6XXX series DC casting, the two main metallurgical issues which are

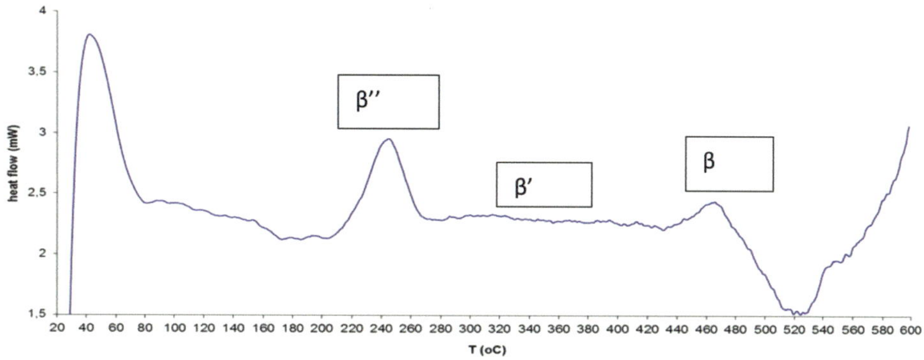

Figure 6.
Typical DSC (Differential Scanning Calorimetry) curves of EN AW 6082 aluminum alloys showing the coherent/ incoherent MgxSiy based intermetallic. (Image copyright © belongs to TRI METALURJI A.S.).

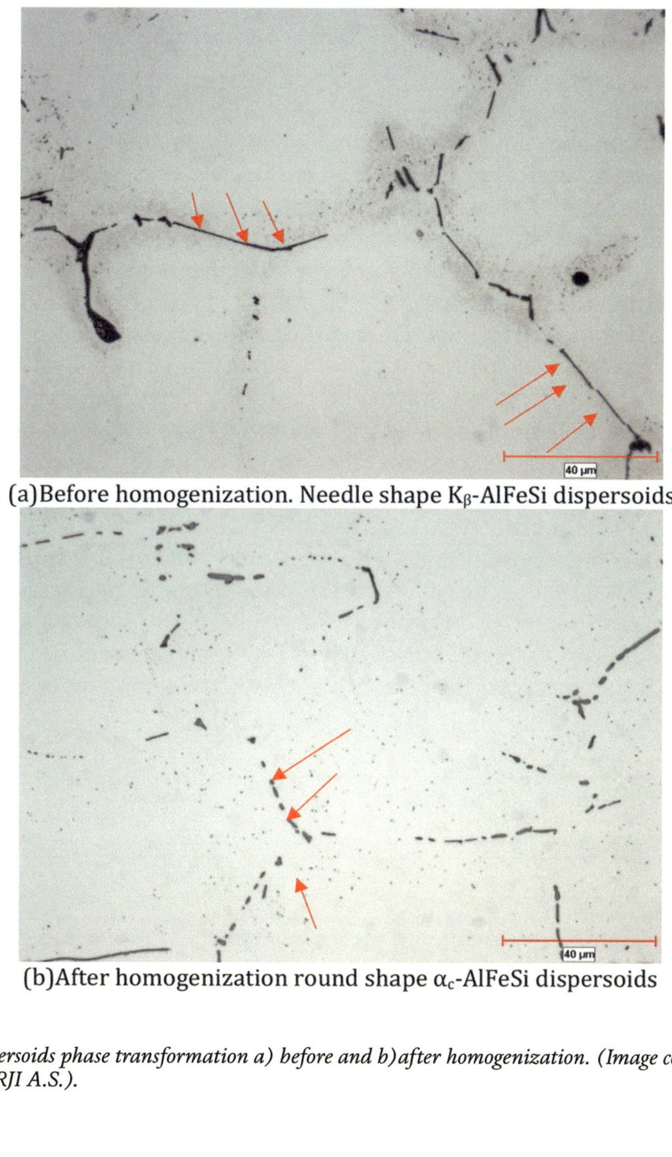

Figure 7.
AlFeSi based dispersoids phase transformation a) before and b)after homogenization. (Image copyright © belongs to TRI METALURJI A.S.).

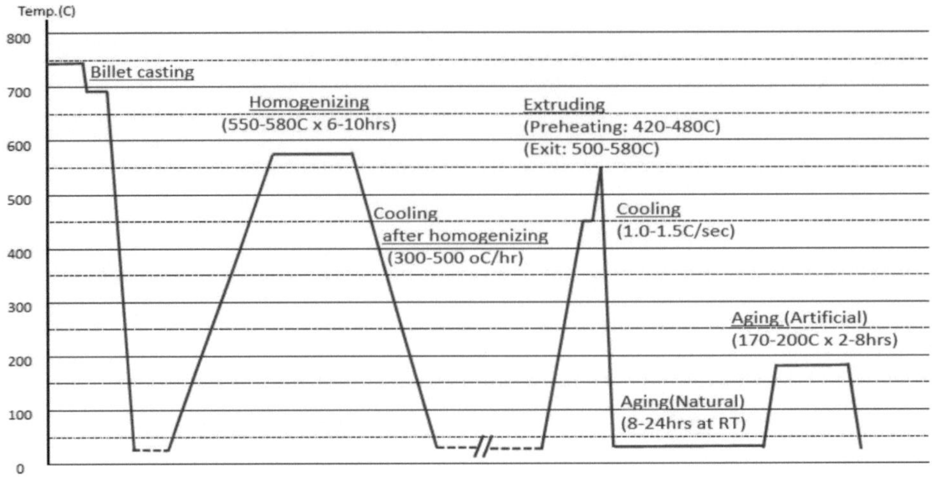

Figure 8.
General thermal history of the 6XXX series aluminum alloys. (Image copyright © belongs to TRI METALURJI A.S.).

MgxSiy precipitation and AlFeSi phase transformation, are being controlled in order to match required mechanical properties and surface effect, respectively.

To understand better of the effect of these two phases, the thermal history of the billet and extrusion process should be analyzed in detail. **Figure 8** shows the thermal history of the 6XXX series aluminum alloys for billet casting and extrusion.

After the casting, the needle shape AlFeSi(Mn,Cr) based dispersoid and very low number of Mg_xSi_y phase are formed. During the high temperature heat treatment, which is called solution heat treatment or homogenization process, all alloying elements starts to diffuse to the aluminum solid solution phases. However, due to the low diffusion rate of some elements such as Fe, Mn, and Cr, the dispersoids are formed but different morphology. **Table 1** shows the diffusion rate (diffusion coefficient rate) of the elements in aluminum at 600°C.

The more important points regarding the diffusion coefficient are that during the solidification of the liquid aluminum alloy, the segregation zone is appearing on the surface of the DC cast billet. The solidification rate on the surface of the billet is so fast that elements cannot move on time to travel to the aluminum solute solution and remains on the dendrite arms and create the inverse segregation zone (ISZ) on the surface. **Figure 9** shows typical ISZ microstructure after homogeneous.

Elements	D/cm^2/s at 600°C	Elements	D/cm^2/s at 600°C
Cu	$4{,}8.10^{-9}$	Fe	$3{,}8.10^{-10}$
Mg	$2{,}1.10^{-8}$	Mn	$2{,}5.10^{-10}$
Zn	$1{,}6.10^{-8}$	Co	$8{,}4.10^{-10}$
Si	$1{,}6.10^{-8}$	Ni	$8{,}1.10^{-10}$
Cr	$1{,}3.10^{-12}$	Zr	$2{,}4.10^{-12}$

Table 1.
Diffusion coefficient of elements in Al at 600°C.

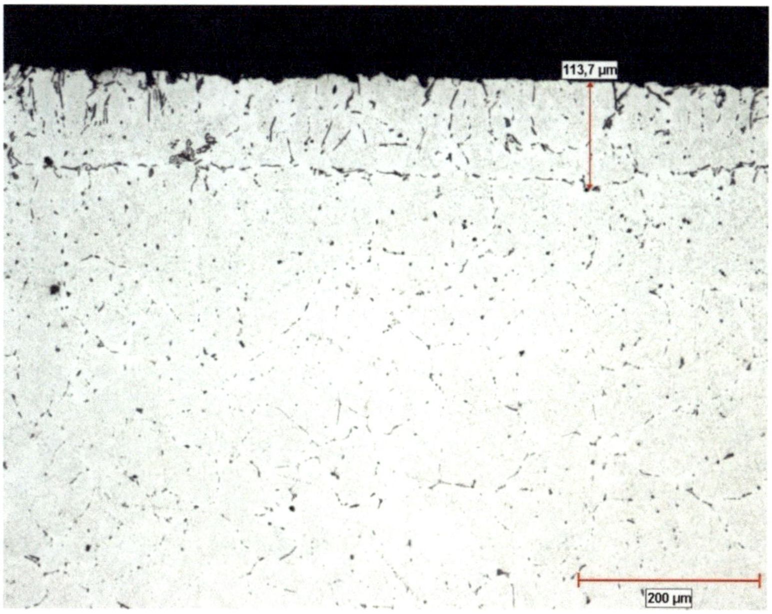

Figure 9.
Typical ISZ formation on the 6063 aluminum alloys. (Image copyright © belongs to TRI METALURJI A.S.).

The Inverse segregation zone is a superficial phenomenon and behaves different depending on casting parameters and alloy chemistry. **Figure 10** shows SEM images of the different surface intermetallic formation on the surface of the billet.

As a result of uneven solidification on the surface, the segregation zone is occurred and, on this zone, the chemical composition is also uneven and different than the bulk material chemical composition. **Figure 11** also shows the uneven element composition (weight%) on the inverse segregation zone area on typical EN AW 6060 aluminum alloys.

During homogenization heat treatment, The needle shape of AlFeSi(Mn,Cr) transforms to rod/round shape AlFeSi(Mn,Cr) phases as shown in **Figure 7**. This transformation is crucial for the following thermo-mechanic process such as extrusion. The needle shape AlFeSi(Mn,Cr) is very brittle and low eutectic temperature behavior and during the extrusion and open to create a surface defect such as pick-up. Besides the surface problem, the productivity of the extrusion (or extrusion speed) will also decrease, and the extrusion die life will be also affected by low phase transformation ratio. On the contrary, the higher amount of transformed AlFeSi (saying as cubic alpha-α_c) will provide higher extrusion speed, better surface condition (or better esthetic) and increase die life. The aim for transformation ratio from needle shape (K_β) to round shape (α_c) should be minimum %95. However, the industrial practices are showing us the average ratio is between %85–95 depending on the furnaces type (batch or continuous type homogeneous furnace).

The second part of the homogenization process is cooling. During soaking period of homogenization process, all Mg and Si atoms dissolve in alpha aluminum phase due their high solubility and create super solute solution structure. The aim of the cooling process is to keep them in solute solution condition. However, it never happens even in high cooling rate. There is always some low percentage coherent β-Mg_2Si present on the system.

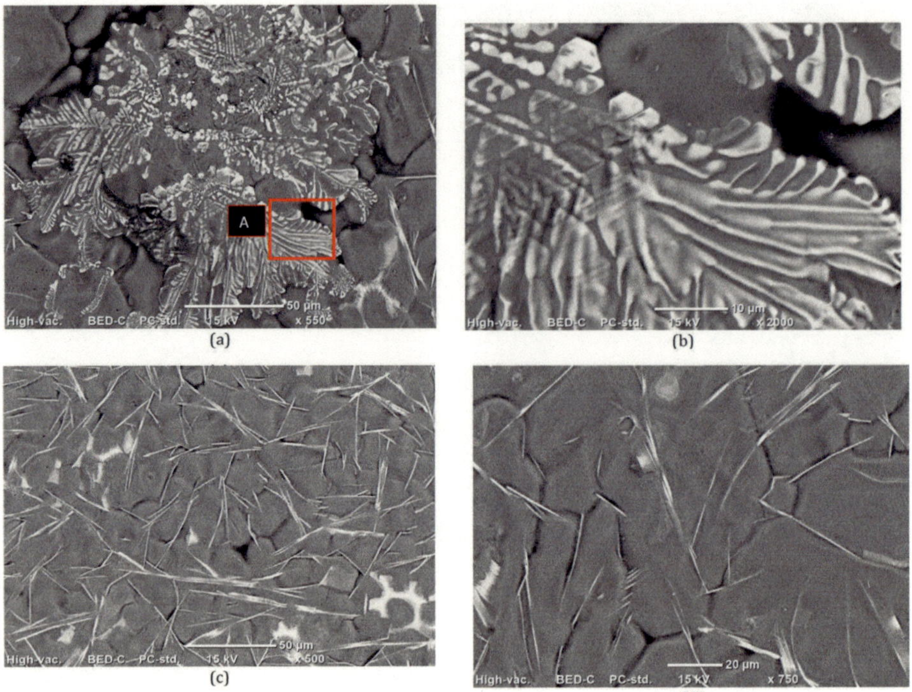

Figure 10.
Different AlFeSi dispersoid morphology formed on the surface of EN AW 6063 aluminum alloys billet during solidification (a) Plate like AlFeSi dispersoid in dendrite structure (b) Plate like AlFeSi dispersoid in dendrite structure (c) Needle shape AlFeSi dispersoid (d) Needle shape AlFeSi dispersoid. (Image copyright © belongs to TRI METALURJI A.S.).

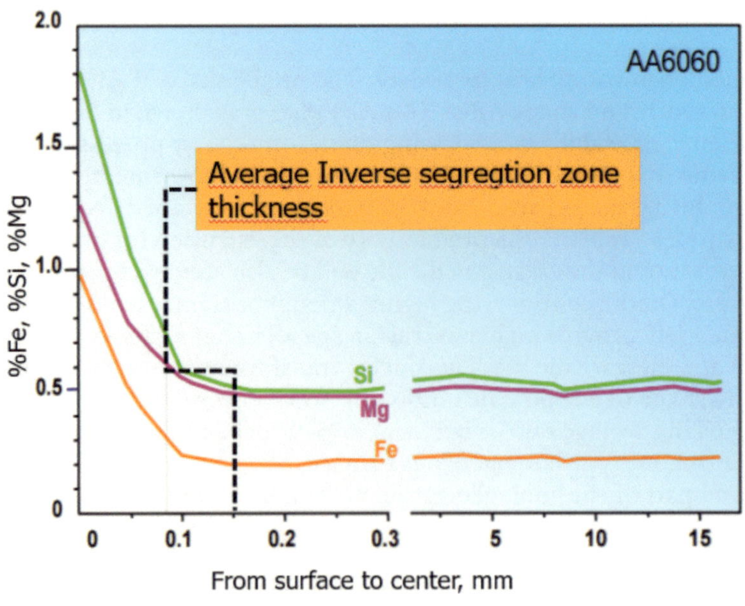

Figure 11.
The element compositional deviation on the inverse segregation zone area [8].

The 6XXX series aluminum alloys are a quench sensitive group and that's why, the quenching system should be well adjusted in billet homogenization and extrusion process as well. **Figure 12** presents the quench effect vs. precipitation of the Mg_xSi_y particles.

As Its clearly seen that on **Figure 12**, the faster cooling provides the more β" phase which is finer and gives better/reliable mechanical properties and faster extrusion speed.

By reducing the cooling rate, especially on EN AW 6082 aluminum alloy, creating the problem on mechanical problems and extrudability of the alloy. The soft cooling practices creates plate like (coarse) Mg_2Si phases (shown in **Figure 3b**) and increase the precipitate free zone. On automotive purposes application such as crash box, the fast cooling is a must to get better folding characteristics. **Figure 13** shows different folding characteristics based on different cooling application after extrusion process.

Copper has appreciable solubility and strengthening effect. The addition of Cu, in variable concentrations, produces substantial solid solution and precipitation strengthening. In presence of magnesium and silicon, copper produces age-hardening effects at room temperature. The weldability and corrosion resistance are decreased, and the weight of the alloy is increased with the copper addition. On 6043 alloys, copper gives more mirror-like surface effect after anodizing process. On the contrary, Increasing the iron level gives more dull surface effect. Copper addition to the 6XXX series aluminum alloys is more popular in automotive purposes profile production such as crash box and some side support profiles which is applied high deformation energy on the car. In case of copper addition to the alloy, the two new phases are forming on the billet microstructure. The first one is Al_2Cu-Θ and the second one $Al_5Cu_2Mg_8Si_6$-Q phase. Theta phase behaves like β (Mg_2Si) phases and depending on the temperature incoherent and coherent phases precipitate and present different behavior. And Q phase is thermodynamically stable phases and enhance the response for long term thermal stability test for crash application. The other important alloying elements are Mn and Cr which give

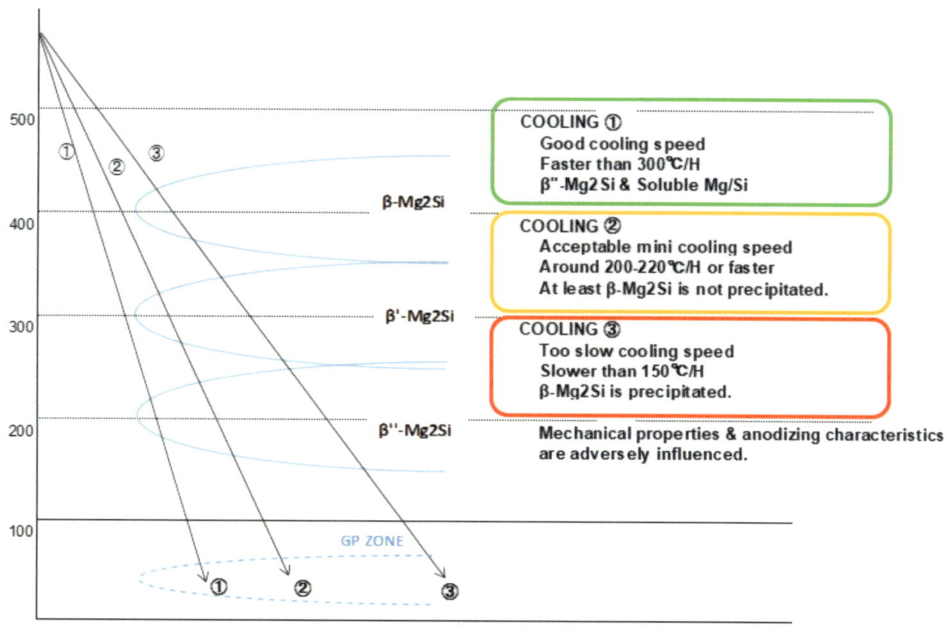

Figure 12.
Precipitation of Mg_2Si vs. Cooling speed after soaking of the billet (Image copyright © belongs to TRI METALURJI A.S.).

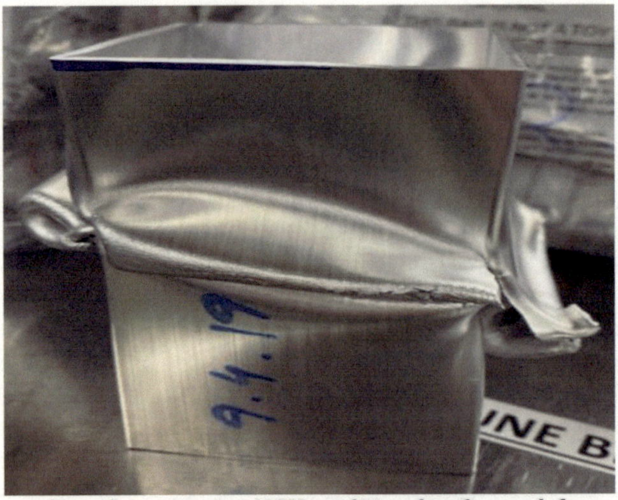

(a) Air cooling after extrusion. NOK condition due the crack formation on folds.

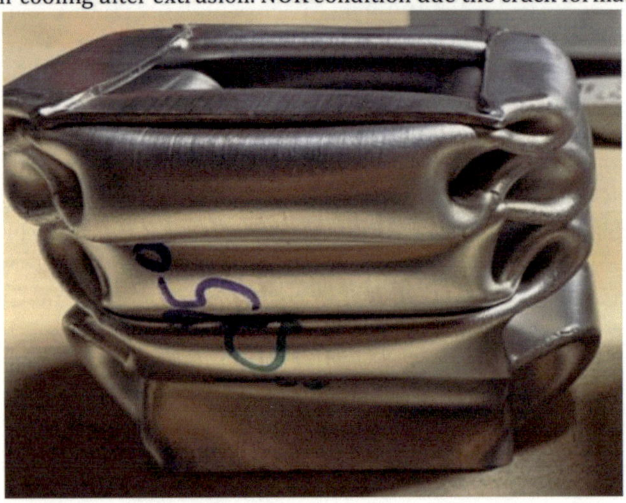

(b) Standing wave cooling after extrusion. OK condition and small crack formation on folds. (OK criteria is defined according to the OEM Company spects)

Figure 13.
Different folding characteristics depending on cooling rate, a) air, b) standing wave (Image copyright © belongs to TRI METALURJI A.S.).

good homogenization response on the 6063 and 6060 aluminum side. And as for 6082 and 6005 A, The Mn and Cr control the grain growth during extrusion and refers to anti-recrystallization elements like other Zr and Sc elements. **Figure 14** shows grain structure micrograph about the grain size control during the extrusion.

Figure 14 shows the effect of Cr element after the extrusion process in grain size. Cr addition creates AlCr based dispersoid and due to grain boundary pinning effect, the final grain size and its structure changed completely from abnormal grain size to equiaxed and fine grain size.

The planned additions of trace elements like Cr, Mn, Zr, and Sc can restrict the softening mechanisms during elevated temperature deformation. The fine grain structure ensures strengthening by grain refinement and work hardening. The trace

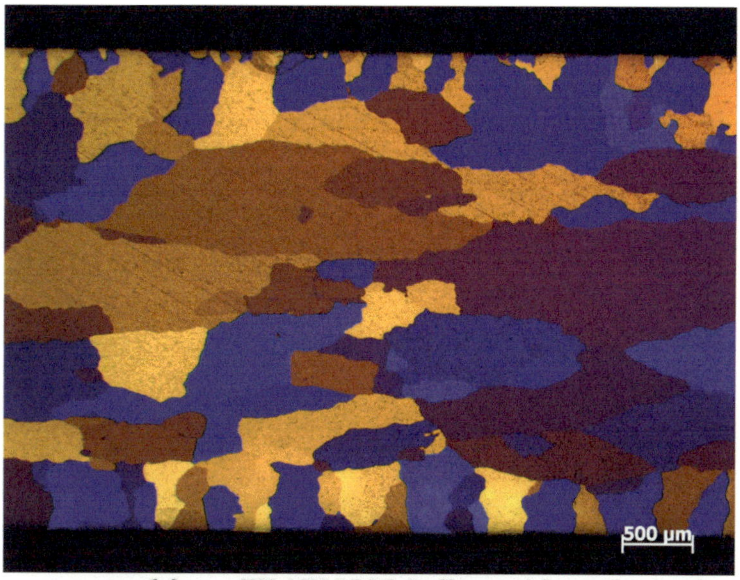

(a) EN AW 6005 A alloys without Cr.

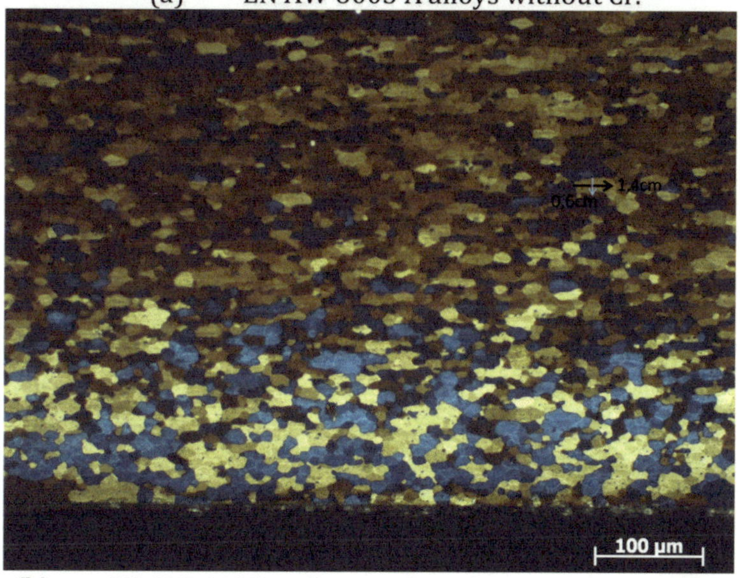

(b) EN AW 6005 A alloys with Cr (0,10-0,15% Cr addition)

Figure 14.
Effect of Cr addition to EN AW 6005 A alloys without Cr and with Cr. (Image copyright © belongs to TRI METALURJI A.S.).

elements show significant role in altering the physical, mechanical, and corrosion behavior of Al and its alloys.

Alloying elements B, Ti in some AA6XXX alloys cause the control of grain size and produce grain boundary strengthening of those alloys. The machinability can be higher by the addition of Bi and Pb in AA6XXX alloys. The solid solution strengthening to some extent can be caused by lattice parameter control in commercial alloys due to presence of Mg and Si in solid solution.

3. Conclusion

Among the aluminum alloys AA6XXX series is the most used alloy group for all engineering applications. Understanding of alloy chemistry and obtaining optimized production parameters and correct usage of materials generates high efficiency with reduced production costs. Even small changes of Si and Mg in alloy chemistry could create differences on the mechanical properties of alloy and combining this alloy chemistry with process parameters of heat treatment shows major progress on the alloy properties and usage.

Author details

Emrah Fahri Ozdogru
TRI METALURJI A.S., Yedpa Ticaret Merkezi, Atasehir-Istanbul, Turkey

*Address all correspondence to: emrah.ozdogru@trimetalurji.com

References

[1] Monoj B, Anil B. Processing and precipitation strengthening of 6xxx series aluminium alloys: A review. International Journal of Material Science. 2020;**1**(1):40-48. DOI: 10.22271/27078221.2020.v1.i1a.10

[2] Demirbolat H, Akdi S, Alkan B. The effect of homogenization and chemical compositions of 6005 and 6082. Aluminium Alloys on The Cold Forming Process. 2021;**28**:16-20. DOI: 10.31590/ejosat.973063

[3] Birol Y. Optimization of homogenization for a low alloyed AlMgSi alloy. Material Characterization. 2013;**80**:69-75. DOI: 10.1016/j.matchar.2013.03.013

[4] Birol Y. Homogenization of direct chill cast AlSi1MgMn billets. International Journal of Materials Research. 2014;**105**(1):75-82. DOI: 10.3139/146.110992

[5] Birol Y. Precipitation during homogenization cooling in AlMgSi alloys. Transaction of Nonferrous Society of China. 2013;**23**:1875-1881. DOI: 10.1016/S1003-6326(13)62672-2

[6] Langerweger J. Effect of metallurgical factors on productivity in the extrusion of aluminum-magnesium-silicon (AlMgSi) alloys. Aluminum. 1982;**58**:107-109. DOI: 10.1080/00084433.2020.1719332

[7] Matias NA, Fernando D. Revision of the solvus limit of Al-Mg_2Si Pseudo binary phase diagram. Procedia Materials Science. 2015;**8**:1079-1088. DOI: 10.1016/j.mspro.2015.04.171

[8] Hauge T. Influence of Al microstructure on hard anodizing quality-profile material. In: Proceeding pf IHAA Symposium. New York; 2014

Section 4

Advanced Applications of Aluminum Alloys

Chapter 8

Conversion of Environmental Waste to Engineering Wealth: Eggshell Particulates as a Reinforcement Agent in Al-6063

Isaac Enesi Dongo and Monday Itopa Momoh

Abstract

In the quest to improve on some properties of aluminum alloy for better engineering application, eggshell particulate was used to reinforce the alloy at various percentages in this research. The reinforcement was carried out at a superheated temperature of 700°C in order to have homogenous mixture prior to casting. Micro hardness, impact and microstructure were evaluated. It was observed that eggshell particulates has positive impact on the mechanical properties of aluminum with sample X_2 reinforced with 4% eggshell particulate possessing the highest hardness value of 187.1 Hv in comparison with the control sample X_o showing the least value of 100.7 Hv. The impact result complemented the hardness results with samples X_1 and X_2 reinforced with 2% and 4% respectively of eggshell particulates exhibiting the highest impact value of 4.7 J each. This thus indicates that eggshell particulate can be considered as an alternative reinforcing agent in improving the mechanical properties of aluminum 6063. The particulate however dampens the corrosion resistance of Al6063 to deterioration in saline environment with the control sample X_o showing the least corrosion rate (0.15711 mpy), corrosion density (-0.012211 A/cm^2) and the highest corrosion potential (0.99186 V). Thus, indicating that the material is not recommended for use in such environment. However, more percentage of the particulates is recommended in order to affirm the optimum ratio to which Al6063 would yield better properties.

Keywords: particulates, Al-6063, micro hardness, corrosion, mechanical properties

1. Introduction

The demand for light weight and cost-effective engineering materials with high strength most especially in automobile, aerospace and structural applications to enhance performance are on the increase. Interestingly, aluminum and its alloy with matching characteristics of light weight, low cost, good thermal and electrical properties desirable in various engineering applications had stimulated research interest. These good properties as well as the recyclability of aluminum had made it a promising candidate for engineering application. Recently, it is used in all sector of human

IntechOpen

endeavor today which is not limited to just structural and automobile applications but also in cooking utensils household furniture, electrical industries, etc., are making use of aluminum either pure or its alloyed form. However, the global increasing demand for light weight material with enhanced mechanical properties over the conventional alloys in the emerging industrial applications and requirement has recently fueled the curiosity of researchers on advanced engineering material (metal matrix composites) developments [1].

Reinforcing a metal matrix phase resulted in metal matrix composite with processing techniques serving as major hub for tailoring the resulting properties into the desired application. Reinforcement material can simply be regarded as strength enhancer or bad carrying member in MMCs. It is usually a non-metallic in nature including ceramic, egg shell waste, agro and bid shell waste [2, 3].

Reinforcement material can simply be regarded as strength enhancer or bad carrying member in MMCs. It is usually a non-metallic in nature including ceramic, egg shell waste, agro and bid shell waste. Metal Matrix Composite (MMCs) was dated back two decades ago. This was to improve physical, mechanical properties and weight serving advantages associated with matrix composite compared to unreinforced alloy [4, 5].

Various viable industrial and agro waste that has been investigated in aluminum alloy matrix include agricultural (rice husk, bamboo leaves, eggshell, animal bone, coconut shell, etc.) waste and industrial (quarry dust) waste material. His investigation shows that agro and industrial waste can be used as aluminum matrix reinforcement material due to their promising properties enhancement [6, 7].

There are various available processing routes for processing of MMCs includes squeeze casting, infiltration powder metallurgy and stir casting technique. However, stir casting process has widely been reported to be most viable and economical route for particle reinforced metal matrix composite (PMMCs) processing and production [8, 9].

Egg shell is an agro-based waste product (after consumption) that is readily available in any society; and its major constituent is calcium carbonate which makes it useful as reinforcement for aluminum 6063. In view of areas of application of aluminum alloys, this research becomes imperative to improve the performance of the material.

2. Experimental procedure

Aluminum alloy 6063 with known chemistry (**Table 1**) was used as base material and eggshell powder used as the reinforcing agent.

The eggshell was sourced from the local tea vendor in Okene, Nigeria. The collected eggshells were rinsed and cleaned thoroughly with fresh water to remove the accompanied dirty and the inner membrane of the shell. It was then dried in sunlight for 7 days to completely remove the moisture. Grinding and pulverization was then carried out to convert the eggshell into powder, this is essential to increase the surface

Element	Al	Si	Fe	Mg	Cu	Mn	Cr	Ni	Zn	Sn
%	89.22	8.36	0.76	0.14	1.19	0.20	0.01	0.05	0.32	0.19

Table 1.
Chemical composition of Al-Mg-Si alloy.

Element	Weight (%)			
	Natural hen eggshell	Boiled hen eggshell	Natural duck eggshell	Boiled duck eggshell
$CaCO_3$	96.48	96.48	96.48	95.99
S	2.31	3.59	1.24	1.92
Mg	0.404	0.440	0.996	0.927
P	0.501	0.469	0.508	0.481
Al	–	–	–	0.309
K	–	–	0.0839	0.00957
Sr	0.0737	0.0734	0.118	0.093

Table 2.
Chemical composition of eggshell.

area of the particulate for effective interaction in due course of melt mixture. The pulverized powder was later fed into the set of vibratory sieve shaker arranged in decreasing order of mesh sizes. This classification yielded a very fine particulate, using 70 mesh size to produce, a particle size equivalent to 0.000212 μm (**Table 2**).

2.1 Preparation of the casting process

In the preparation of the casting process, charcoal, crucible pots and molding sand was sourced. This process involves the use of molten material usually metal poured into a mould cavity that takes the form of the finished parts. The mold material then cools with heat generally being extracted via the mold, until it solidifies into the desired shape.

A plastic pattern of cylindrical shape diameter of 6 mm and length of 30 mm was used. Adequate shrinkage allowance was provided. Silica sand was gotten and mixed manually with bentonite (as a binder) and water to ensure mouldability.

2.2 Specimen preparation and analysis procedure

The furnace was charged and fired to a superheated temperature of about 700°C. While waiting for the furnace to increase in temperature the eggshell was being weighed on the weighing balance in grams gotten from the charge calculation representing 2%, 4% and 6%. The Al-6063 alloy was charged into the crucible pot and placed in the furnace. The alloy turned liquidus at a temperature of 660°C but was left to super heat so as not to turn solidify upon removal. 2% eggshell particulate, as weighed, was added to the melt and stirred for about 60 s and the combined charge was left to stay in the furnace for about 10–15 min before it was removed and poured inside the mould for solidification. This process was repeated for 4% and 6% eggshell particulates. A control sample was also cast without reinforcement.

The first casting was labeled X_0 representing Al-6063 without reinforcement. This serves as the control sample; X_1 representing 2% of eggshell particulate; X_2 representing 4% of eggshell particulate; and sample labeled X_3 representing 6% of eggshell particulate (**Table 3**).

The hardness of the sample was evaluated using vicker's micro hardness testing machine to determine its ability to resist indentation when subjected to load of 300 g

Label	Weight of materials (kg)	% weight of particulate	Equivalent weight of particulate (g)
X_o	2.26	–	–
X_1	2.26	2	45.2
X_2	2.26	4	90.4
X_3	2.26	6	135.6

Table 3.
Particulate mixture in Al-6063.

for a dwell time of 10 s using standard procedure (ASTM E-384-17). As a complement, Impact testing of metals was performed to determine the toughness of both reinforced and unreinforced Al-6063 by calculating the amount of energy absorbed during fracture (ASTM A-370).

The surface of the developed composite was examined using a scanning electron microscope (SEM). The scanning electron microscope uses a focused beam of high energy electron to penetrate a variety of signals at the surface solid specimen. The signal that is derived from the electron sample interactions was reveal the information about the sample including the external morphology, texture, chemical composition, crystalline structure and the orientation of the material that makes up the samples.

3. Results and discussion

3.1 Effect of eggshell particulate on the hardness

Figure 1 shows the average hardness plot of the hardness of reinforced Al6063 with eggshell. Micro hardness testing machine was used to evaluate this property. In the course of conducting the test, a test load of 4.904 N was used over a dwell time of 10 s. Three indentations were made on each sample and the average calculated and documented. From the result, it was observed that there was a general increment in the hardness values of the reinforced samples. While there were gradual but rapid increase in the hardness value to the peak in sample X_2 reinforced with 4% eggshell particulates, a drop was observed in the X_3 sample (reinforced with 6% eggshell particulates). Generally, the reference sample was observed to display the least hardness value and the reinforced sample indicates that eggshell particulates are of much importance to Al alloy for better engineering performance where high strength is required. The excel model shows that the optimum hardness value of the composite can be attained at about 4.6% eggshell particulate reinforcement with a rank correlation coefficient of 1.

3.2 Effect of eggshell particulate on the impact energy

Like the hardness experiment, impact test was conducted at a room temperature of 25°C. The experiment was necessary in order to evaluate the toughness or the amount of energy the reinforced samples can absorbed prior to failure. From the result in **Figure 2**, it was observed that samples X_1 and X_2 reinforced with 2% and 4% of eggshell particulates exhibited the highest impact value with a sharp drop in sample X_3

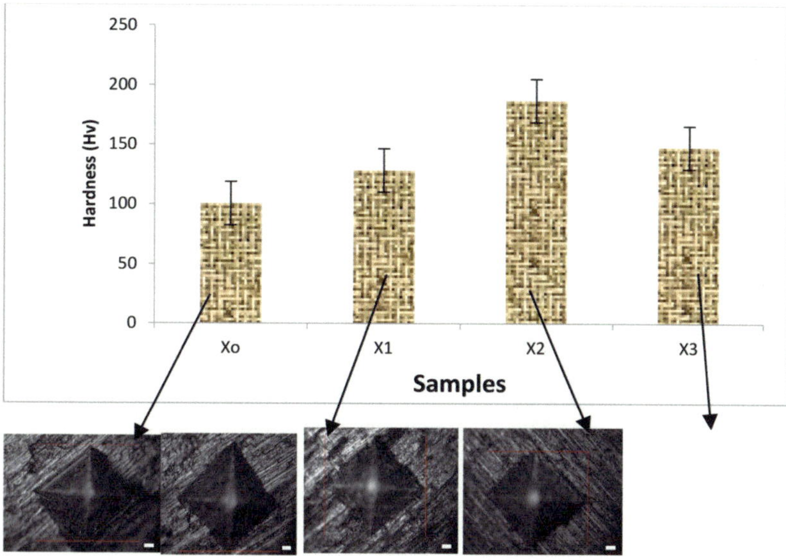

Figure 1.
Hardness trend showing the effect of eggshell particulates on Al-Mg-Si alloy.

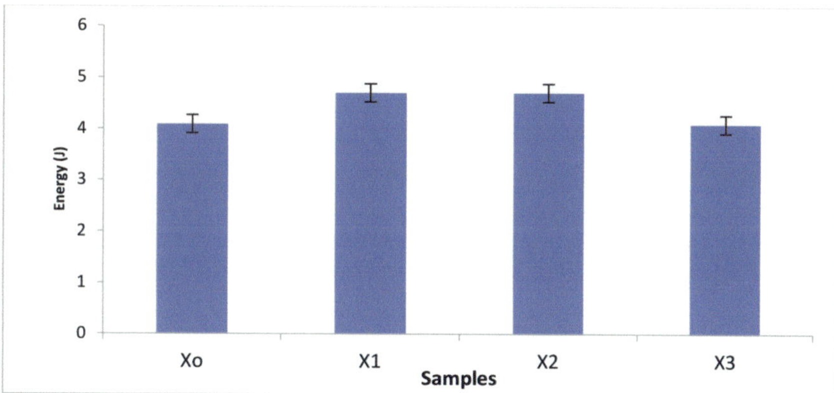

Figure 2.
Toughness of reinforced Al-Mg-Si alloy.

reinforced with 6%. This thus implies that Sample X_1 and X_2 displays a positive effect of the particulates and can be useful in a vibratory environment within the scope of other properties considered.

3.3 Effects of eggshell on the microstructure

Figure 3(a–d) shows the micro view of the structure of the aluminum alloy (both reinforced and as-cast). The image was captured at high resolution of 1000×, the resolution was ensured to be consistent in order to juxtapose the multi-images. From the displays, it shows that there was a metal flow in the sample X_o reinforced (without reinforcement), the flow could be due to the absence of hardness enhancer. This limitation is taken care of in the reinforced sample (X_1, X_2 and X_3) which shows a display of the eggshell particulates precipitates distributed round the structure of the material. This precipitates account for the significantly improved mechanical properties of the reinforced alloy under consideration. Also, the uniformly distributed particulate in the structure shows that there was thorough stirring, thus homogeneity, of the melt prior to casting.

3.4 Effects of eggshell particulates on the corrosion behavior

Machine designed to operate in marine environment are generally vulnerable to failure by corrosion. It is therefore imperative that the corrosion susceptibility of the materials used for such facility be investigated to establish the mechanism for its mitigation. The unreinforced and reinforced samples were subjected to potentio-dynamic electrochemical measurement technique in accordance to ASTM G34 standard to determine its behavior in simulated saline environment.

Figure 4(a–c) show the extracted parameters from the linear polarization plots, results shows that the reinforced alloy generally displayed similar polarization curves and passivity characteristics. It was observed that in spite of the recorded improvement in the mechanical properties, the entire reinforced sample exhibited poor corrosion resistance in saline environment. The as-cast sample (without reinforcement)

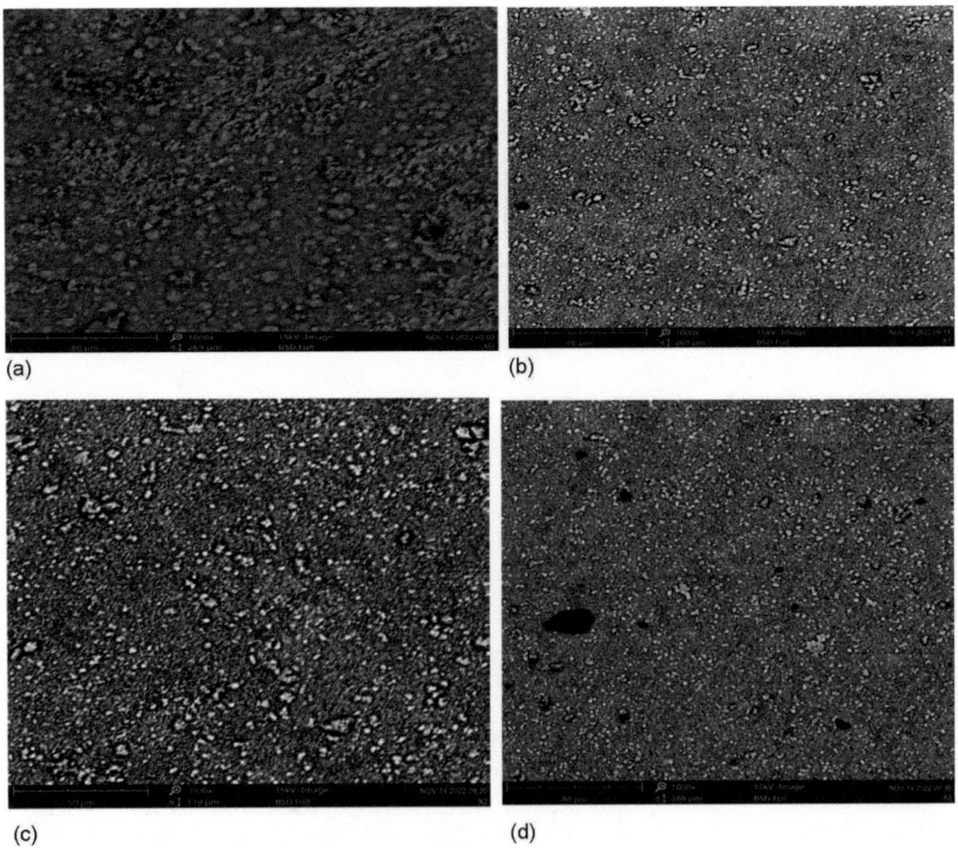

Figure 3.
(a) SEM view of as-cast Al-Mg-Si alloy, (b) SEM view of sample X_1, (c) SEM view of sample X_2 and (d) SEM view of sample X_3.

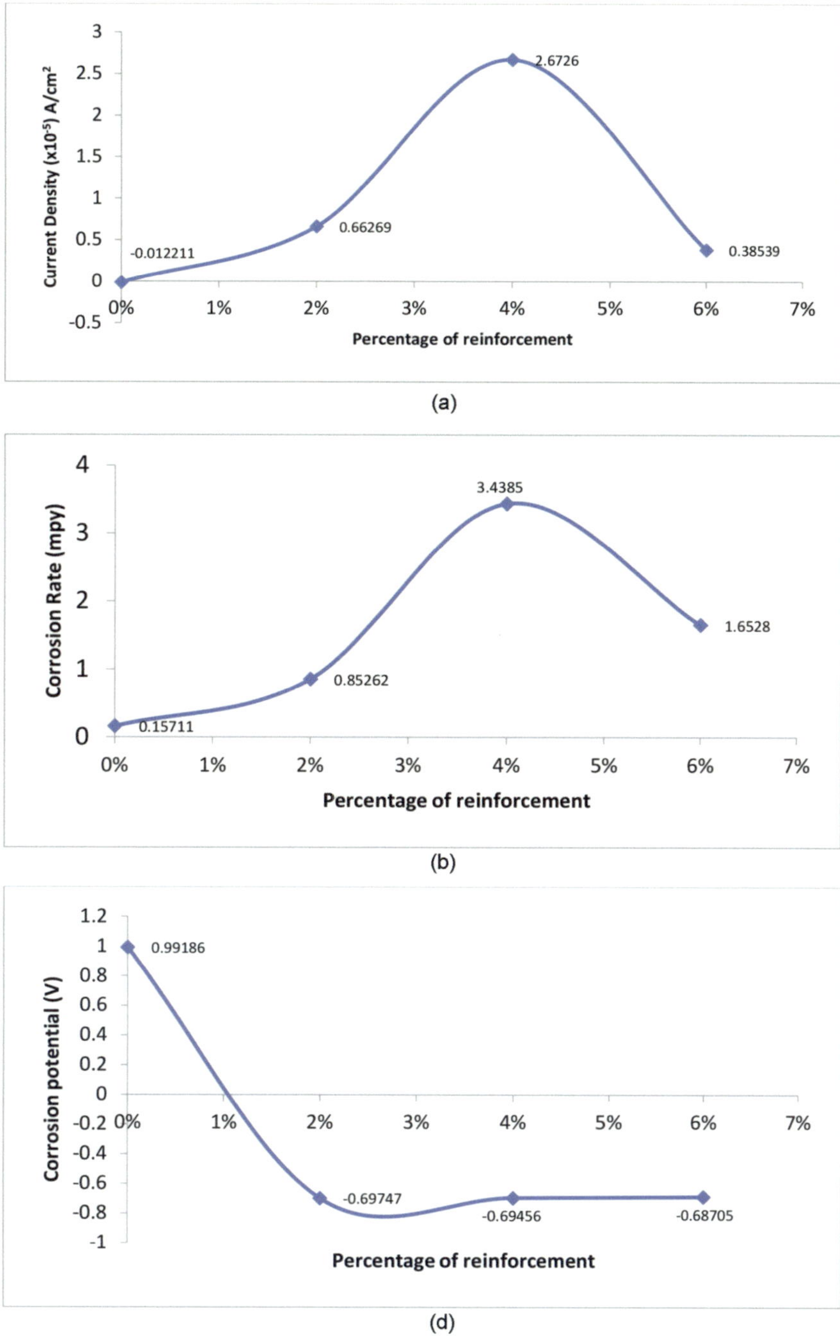

Figure 4.
(a) Showing the current densities of the samples, (b) showing the corrosion rates of the samples, and (c) showing the corrosion potentials of the samples.

has the lowest corrosion rate of 0.15711 mpy compare with those reinforced with 2.0 wt.% eggshell (0.85262 mpy), 4.0 wt. % eggshell (3.4385 mpy), and 6 wt.% eggshell particulate (1.6528 mpy). Correspondingly, the unreinforced sample has the least corrosion current density of -0.01221×10^{-5} A/cm^2 and the highest corrosion potential of 0.99186 V compared with the reinforced sample exhibiting high current densities and potentials. From the above corrosion parameters obtained from the electrochemical test, it was discovered that the unreinforced Al6063 has the best corrosion resistance hence it has the least corrosion current density, lowest corrosion rate and the highest corrosion potential.

4. Conclusion

The mechanical properties of Al-6063 have been enhanced in this research owing to the adoption of eggshell particulates as reinforcing agent. The reinforcement was carried out at varied percentages and at a superheated temperature of 700°C in order to ensure homogeneity in the mixture prior to casting. Micro hardness, impact and the microstructural characterization were carried out under varied condition. From the results, it was observed that eggshell particulates have positive impact on the mechanical properties of Aluminum under consideration. While a significant rise in the hardness value was observed, the gradual increment in the impact result complement the hardness results. This thus indicates that eggshell, which ordinarily is considered a dangerous agro-waste, can be considered as an alternative reinforcing agent in improving the mechanical properties of Aluminum 6063. The particulate however dampens the corrosion resistance of Al6063 to deterioration in saline environment, indicating that the material is not recommended for use in such environment.

Author details

Isaac Enesi Dongo[1*] and Monday Itopa Momoh[2]

1 Department of Metallurgical and Materials Engineering, Kogi State Polytechnic, Lokoja, Nigeria

2 Department of Foundry Engineering Technology, Kogi State Polytechnic, Lokoja, Nigeria

*Address all correspondence to: isaacdongo@gmail.com

References

[1] Aweda JO, Kolawole MY, Sulaiman A. Archachatina Maginata bio-shell as reinforcement material in metal matrix composite. International Journal of Automobile and Mechanical Engineering. ISSN 2180-1606. 2017;**14**(1):4068-4079

[2] Bahram A, Soltani N, Pech-Canul MI, Gutierrez CA. Development of metal matrix composite from industrial/agricultural waste material and their derivatives—A review. Critical Review in Environmental Science and Technology. 2015;**46**(2):143-208

[3] Akbar HI, Surojo E, Ariawan D. Investigation of industrial and agro waste of Aluminum matrix composite reinforcement. Pocedia Structural Integrity. 2020;**27**:30-37

[4] Nithyanandhan T, Ramamoorthi R, Sivasurya KV, Vesvanth M, Sanjay S, Sriram K. Machining behaviour of Aluminium 6063 with strengthen aluminium oxide and chicken bone ash produced by modern technique. Journal of Science Technology and Research. 2021;**2**(1):11-23

[5] Karthikeyan K, Dhinakaran V, Rajkumar V. Evaluation of mechanical properties of aluminium 6063—Borosilicate reinforced metal matrix composite. AIP Conference Proceedings. 2020;**2283**(1):1-6

[6] Joseph DO, Anoja J. Characterization of Al-Mg-Si Alloy Reinforced with Eggshell Particulates. An unpublished project. Nigeria: Kogi State Polytechnic; 2022. p. 37

[7] Alaneme KK, Adegun MH, Archibong AG, Okotete EA. Mechanical and wear behaviour of aluminium hybrid composites reinforced with varied aggregates of alumina and quarry dust. Journal of Chemical Technology and Metallurgy. 2019;**54**(6):1361-1370

[8] Garba MJ, Shehu UY, Ejeh SP, Lawal A. Combined effects of coconut shell-groundnut shell ash (CSA-GSA) on the properties of concrete. Journal of Research Information in Civil Engineering. 2019;**16**(3):2816-2822

[9] Edoziuno FO, Nwaeju CC, Adediran AA, Odoni BU, Arun Prakash VR. Mechanical and microstructural characteristics of Aluminium 6063 Alloy/Palm Kernel shell composites for lightweight applications. Scientific African. 2021;**12**:1-11

Chapter 9

The Microstructure and Mechanical Properties of the Aluminum Alloy (AA 6061 T6) under the Effect of Friction Stirs Processing

Emad Toma Bane Karash and Mohammad Takey Elias Kassim

Abstract

The following chapter study the friction stir processes (FSP) is used to improve the surface characteristics of the alloy AA6061-T6 on the surface topography, hardness, tension mechanical characteristics, and microstructures of aluminum alloy, the impacts of friction stir process tool travel and rotation speeds were investigated. All friction stir processes (FSW) in this investigation used a cylindrical tool without a pin that had a 20 mm diameter, rotated at different rotating speeds 800, 1000, 1250, and 1600 rpm, and at different travel speeds 32, 63, and 80 mm per minute. The examination of the current study's data and the test results showed that in stir friction processes, hardness rises with cutting depth. The study of the crystal structure showed that the hardness increased by twice as much for two stages as it did for one stage. Additionally, it was observed that as cutting depth increased, the size of the granules representing engineering defects grew smaller. Additionally, in the case of two stages, the ratio of granule size to friction was twice as high as in the case of one step. According to the results, using a single-stage friction stir process increased yield strength by 18% and tensile strength by 9.5%, while using a two-stage friction stir process increased yield strength by 20.4% and tensile strength by 11.5% when compared to metal basis.

Keywords: friction stir processes, aluminum alloy, linear speed, travel speed, hardness, microstructure

1. Introduction

The lightness of aluminum in comparison to other forms of metal is one of its qualities that led to its widespread application in all industrial sectors. Due to its density, which is around one-third that of steel as an engineering material, and the fact that heat-treatable aluminum alloys may provide a high percentage of resistance, aluminum ranks second in importance only to steel as an engineering material. It ranks third among the earth's accessible elements, behind silicon and oxygen, making it one of the most abundant metals in the crust of the planet.

It also happens to be one of the most chemically efficient metals, making up 7.5% to 8.1% of the weight of the earth's surface [1, 2]. The ability of heat-treated aluminum alloys to withstand corrosion by water, salts, and some other substances makes this metal resistant to natural climate variables, which is another significant characteristic in the business. Aluminum's high affinity for the union with oxygen and the development of a thin surface layer are what cause it to have such a good ability to resist corrosion. From its oxide, it is impenetrable (Al_2O_3). The metal is exposed to air, which causes it to develop. Corrosion is halted by it. As a result, it is used to make tanks for the chemical industry, as well as to make colors and use its powder in paint. An additional benefit of aluminum is that it is a non-toxic metal, making it a valuable material. Corrosion is prevented by it. As a result, it is employed in the creation of tanks for the chemical industry as well as the production of colors and paint using its powder. Aluminum is utilized in the creation of containers for storing food and soft drinks since it is a non-toxic metal in addition to having all these other significant qualities [2, 3]. Numerous previous researchers have examined and investigated the friction stir welding processing parameters for improving the mechanical qualities of the welded joint [4–6]. It would be equivalent to claiming that different joint technology alloys are employed in fabrication because fusion welding has melted a lot of them while producing secondary phases due to thermal absences using base materials as a pin device [7]. Due to this, compared to alloys of various compositions, the temperature distribution in nugget zones (NZs) has risen more gradually. The best solution to this issue is to gather rotational and transverse speeds. The tool is therefore essential in FSW [8]. It has been addressed in detail how to fabricate surface composites with different reinforcement combinations, such as AA7075-TiO_2, AA7075-and-TiO_2-SiC, and TiO_2 + SiC, while maintaining consistent tool rotation, tool travel speed, and number of passes. The same is meant to increase hardness. It has been noted that the friction stir treated surface composite had an average hardness that was 4 times more than the parent metal [9, 10].

2. Friction stir process studies

The technique of friction and stir treatment derives from surface engineering technology, which allows localized removal of plumbing flaws by changing or smoothing the microstructure and enhancing the material's mechanical properties. The investigation of the friction stir process (FSP) for aluminum alloys was the subject of a number of studies.

J. M. Salman [11] investigated the impact of two rotational speeds (710 rpm) and three linear speeds (116, 189, and 303 mm/min) on the mechanical characteristics and microstructure of an aluminum-zinc-magnesium-copper alloy. Shock resistance, elongation, a rise in the hardness value, and an increase in the tensile and yield strength were all seen to increase during the process. Observations were made along the original and treated samples' structural details in great detail. A variety of thermal treatments, including fermentation, equation, and hardness by quenching in water, were applied to several samples that shown improved fermentation process parameters. The outcomes also demonstrated an increase in the heat produced during the procedure. The findings also revealed a drop in the linear speed and an increase in the heat produced throughout the operation when the rotational speed was raised.

A friction stir procedure was used to blend SiC particles with an aluminum alloy type (6351) to create molecular composite materials in a work by V. V. Murali Krishna et al. in 2015 [12]. Using a friction stir process tool with or without a motor, the samples were rotated and lateral accelerated at constant rates. An optical microscope was used to obtain the (SiC) reinforcements and make measurements of the microstructure of the changed surfaces. At the University of Technology Malaysia, the mechanical characteristics were estimated using the tensile test. The study demonstrated that the number of multi-stage runs results in an improvement in the crystal structure of the friction stir process, and the results of the tensile tests showed an increase in the toughness and hardness of the microstructure for one run of the friction stir process of the models, but a decrease in the durability of the models with the increase in runs, which is attributed to the dissolution of solid precipitates.

The effects of the frictional stir process on the structural and mechanical properties of an aluminum casting alloy with a high degree of hardness were studied by Karthikeyan et al. in 2008 [13], using three different feeding rates (10, 12 and 15 mm/min) and two different tool speeds (1400 and 1800 rpm). It was found that increasing the tool speed for a given feed results in an improvement in the mechanical properties. Additionally, the hardness values of the internal structure of the processed materials rise as a result of the machine's increased rotational speed. As evidenced by the data, the internal structure's hardness values become as close as feasible to the internal structural hardness of the primary materials as one moves away from the weld mass's center. The findings of the mechanical properties process also reveal optimal values at a feed rate of (12 mm/min). It was noted that the average crystal size of (350 nm) in models worked using a machine with a horizontal linear speed of (15 mm/min) and a rotational speed of (1400 rpm) in When the average crystal size was 400 nm) for the same feeding speed and rotational speed. The average crystal size in (25) places on the base material (6.2 m), and it was noted that the average crystal size of (350 nm) in models worked using a (1400 rpm).

In terms of the study, Al-Danaf et al., 2010 [14] examined how the frictional stir process affected commercially rolled sheets (AA5083). A rotary tool with a rotating speed of 430 rpm and a transverse feeding rate of 90 mm/min was used to apply friction stir to the models. And that after this process, a 1.6 m-wide fine crystalline microstructure was obtained. It was discovered that a reduction in the crystal size resulted in a noticeably improved level of ductility and a reduction in forming forces. The ductility was determined using the tensile elongation at a temperature of 250°C, at three stress ratios. The frictional stir procedure increased the ductility of the working materials with a coefficient that varied between (2.6–5) when compared to the ductility of the final product, in the same range of stress rates that were examined. The original metal's Vickers hardness test score was (HV 80), however the number at the center of mass was higher and reached roughly (HV 95).

3. Metal property

Aluminum alloy (T6) - 1 (AA606) heat-treated sheets were employed as the foundation metal in this work. **Table 1** show the alloy's chemical makeup, and **Table 2** that can be trusted provided the alloy's mechanical characteristics. According to Table, this alloy is allowed for use with aluminum alloy [15–18].

Element Materials (%)	Nominal chemical composition	Actual chemical composition
Zn	Max. - 0.15	0.108
Si	0.4–0.8	0.415
Fe	Max. - 0.08	0.408
Ga	—	0.011
Cu	0.15–0.4	0.269
Mn	Max. - 0.15	0.063
Mg	0.8–1.2	0.895
Cr	0.04–0.35	0.208
Ni	—	0.013
Ti	Max. - 0.15	0.017
V	—	0.011
Others each	Max. - 0.05	0.005
Others total	Max. - 0.15	0.025
Al	95.8–98.6	97.552

Table 1.
Chemical composition of AA 6061-T6.

Material properties	Nominal value [15]	Actual value
Tensile Yield Strength MPa	276	245
Ultimate Tensile Strength MPa	310	309
% EL	17	16
Modulus of Elasticity Gpa	69.8	68
Hardness, Vickers	107	101.5
Hardness, Brinell	95	93.5

Table 2.
Mechanical properties of AA 6061-T6.

4. Initialization of friction stir process (FSP)

Aluminum alloy (AA 6061-T6) with a nominal composition in (Wt. %) was the substance used. The dimensions and thickness of the ingot were adjusted on a milling machine to obtain the necessary dimensions (220 * 30 * 3 mm in size), after which it was prepared in a different way (Butt joint configuration) so that the welding line is perpendicular to the rolling direction. This was done to prevent altering the boards' original composition. Using a set rotational speed for the friction tool and a welding linear speed of 800 rpm and 80 mm/min, the friction stir process was performed on a typical vertical milling machine. In both stages, the friction tool is at varying depths of 25, 50, 100, 150, 200, 300, and 400 m. A single pass of the friction tool was made during the first stage of the samples' friction process, and a double pass was made during the second stage on a plate of thickness (3 mm). **Figure 1** depicts the friction model's dimensions and shape.

The friction stirrer is typically made of carbon tool steel (X12M), which has undergone heat treatment to achieve the desired hardness (62.85 HRC), giving it a high degree of durability and resistance to heat and sulfating. The chemical components of the friction stirrer have also been examined in the Department of Standardization and Quality Control in Baghdad; further information is provided in Appendix (B). The friction tool's diameter was concave (mm 10). The friction tool's geometry, as indicated in **Figure 2**, was utilized. The chemical components of the friction stir process tool utilized are described in full in **Table 3**, which was chosen in accordance with **Table 4**.

All friction stir activities in the mechanical workshop are performed on the vertical milling machine. We also created a straightforward fixture out of carbon steel to hold the components while they were being welded together and fixed using the friction technique. In the present study, four rotational speeds (800, 1000, 1250, and 1600 rpm) and three linear speeds (32, 63, and 80 mm/min) were selected from the factors of frictional stir processes, namely the linear speed of the milling machine table and the rotational speed of the friction tool. To complete the friction stir procedure, the tool was turned clockwise (C.W.). Some models of aluminum alloy and the procedure for carrying out the friction stir process are shown in **Figure 3**.

In order to determine how the frictional stir process affects the alloy's properties, frictional stir operations are carried out on the alloy for one stage (one pass) and two stages (double pass) at various depths (**Figure 4**).

For both the first stage and the second stage of the models, light microscopy and scanning electron microscopy were used to analyze the microscopic structure of the friction stir process models. Where I exposed the grain boundaries using Keller's reagent (for a duration of 10–20 sec.) at room temperature and for all samples. The friction zone hardness was measured by applying a load of (27 gm.) for (15 s) using an indenter) of a pyramid shape made of diamonds according to the specification (ASTM-E384) [20]. The precise hardness measurements were performed using a Vicker's micro hardness device (Zwick/Roell Z HV). As shown in **Figure 5**, readings are obtained at three sites for each model, one of which is in the center of the frictional

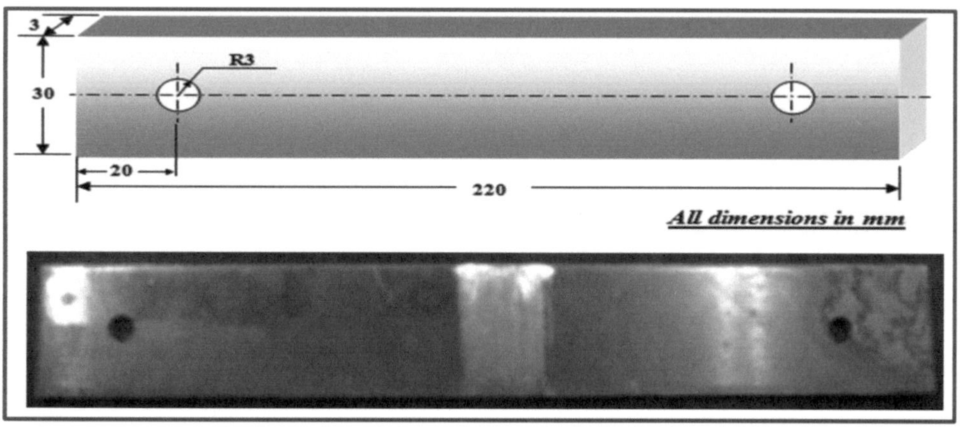

Figure 1.
Displays the size and shape of the alloy friction stir process model (AA6061-T6).

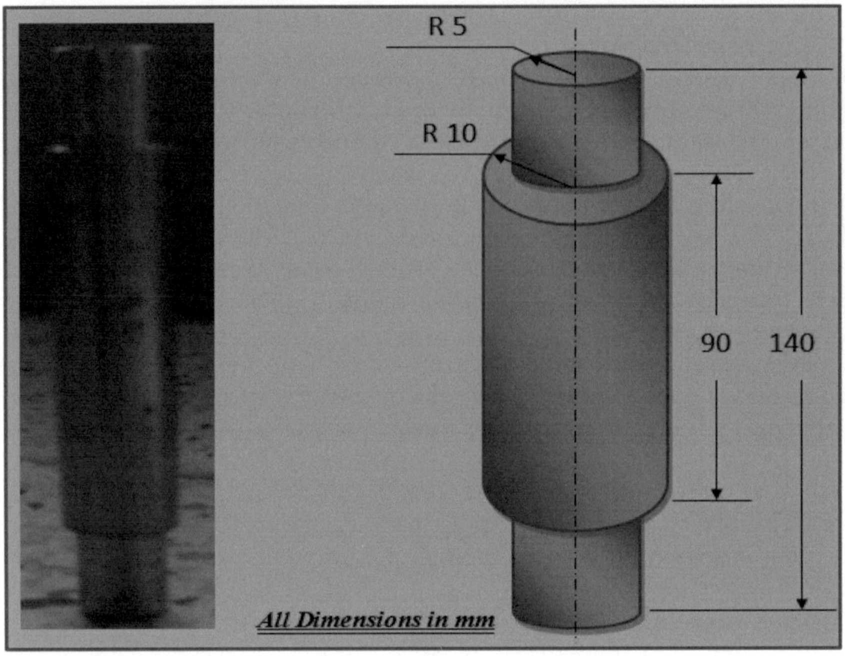

Figure 2.
The tool's geometric design, as used in the friction stir procedure.

Model	Specification	Mechanical tests			
Alloy Steel	ISO 6508	Hardness (HRC)		62.85	
		Chemical composition			
		C %	0.845	Al %	0.007
		Si %	0.265	Co %	0.098
		Mn %	0.240	Cu %	0.118
		P %	0.037	Nb %	0.001
		S %	0.009	Ti %	0.006
		Cr %	4.17	V	1.79
		Mo %	2.31	W	6.52
		Ni %	0.136	Sn %	0.005
		Fe %		Rem.	

Table 3.
Lists the chemical components of the tool used in the friction stir operation.

stir area close to the surface. The other two readings are taken on a straight line to the right and left of the center, with a 1 mm gap between them (**Figure 5**). With a rotational speed of 800 rpm and a linear speed of 80 mm/min and different cutting depths (25, 50, 100, 150, 200, 300, 400 m), transverse tensile models for the base metal and friction tool stir process were prepared in two stages, with the first stage being once and the second stage being twice.

		Thickness	
No.	**Alloy**	**mm**	**Tool material**
1	Aluminum alloys	< 12	Tool steel, WC – Co
2	Magnesium alloys	< 6	Tool steel, WC
3	Copper	< 50	Nickel alloys, PCBN (a)
4	Copper alloys	< 11	Tool steel
5	Titanium alloys	< 6	Tungsten alloys
6	Stainless steels	< 6	PCBM, tungsten alloys
7	Low – alloy steels	< 10	WC, PCBN
8	Nickel alloys	< 6	PCBN

a. PCBN, polycrystalline cubic boron nitride

Table 4.
It displays a few of the materials used to make welding tools [19].

Figure 3.
Frictional stir process of models.

5. Macro and microstructure specimens

Following the friction stir and welding processes, the samples are prepared for microscopic inspection using the following steps:

Samples were created from acrylic material using hot supports after being cut to length and thickness from the cross-section of the area to be investigated. Size and fineness of hydrate (graded from coarse to fine (220, 320, 600, 1000, 1500, 2000 grit). The samples are then polished and polished using diamond paste with a granular

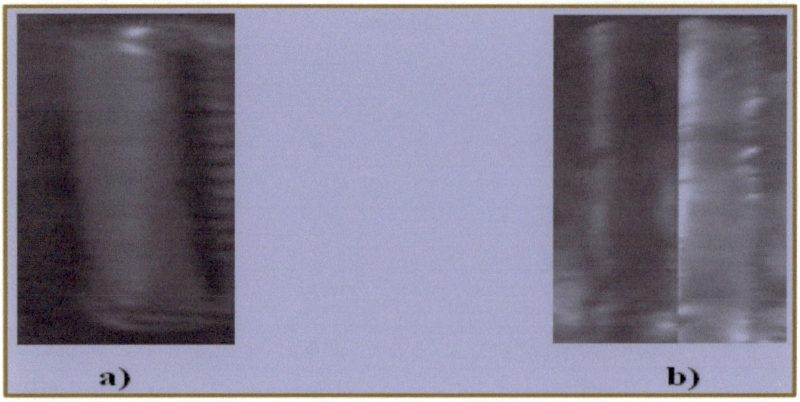

Figure 4.
Multiple stages of friction stir processes for models: A-single stage, b- two stage.

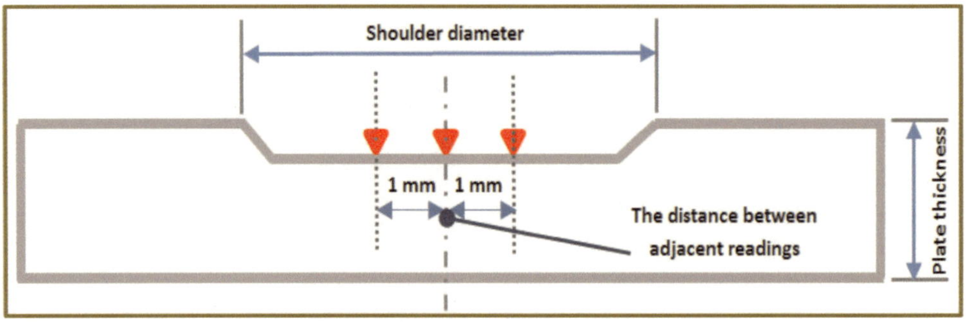

Figure 5.
Shows how to perform an accurate hardness reading.

size of m 1 and lubricating oil to produce a bright, scratch-free surface that resembles a mirror. In addition to water grinding, the sample may also be submerged in alcohol for a brief period of time, carried with specialized forceps, and then washed with running water after the polishing and polishing procedure. Finally, the samples undergo an exposition procedure (also known as etching) using Keller's reagent solution for 10–20 seconds. Following this, the samples are rinsed in water, their surfaces are cleaned with methanol alcohol, and they are dried by being placed in heated air. According to ASTM E3–01 [13], the microstructural investigation was performed on all samples at room temperature using the following materials: distilled water, concentrated hydrochloric acid (1.5 ml), concentrated nitric acid (2.5 ml), and concentrated fluoric acid (1.0 ml) (95 ml).

A magnification light microscope (X20) was used to prepare the samples for microscopic analysis, and the microstructure of the base metal and the other samples were discovered.

6. Friction stir processes (FSP)

The metallurgical and mechanical properties of the stir area change as a result of the heat produced by friction between the stir tool and the components to be welded and the plastic deformation in the stir area during the stir process (FSP). A number of operational and engineering elements, such as the friction tool's rotating speed, the table's linear speed, the stir device's form and shape, the quantity of heat entering the stir region, and other considerations, have an impact on these changes.

6.1 The best model to use for the friction stir procedure

The sample of the optimum model, which achieves the best mechanical properties within the best suited parameters, was tested at the friction tool's rotating speed of 800 rpm and linear speed of 80 mm/min for one stage and a depth of 400 m, as indicated in **Table 5**. The average value of the modulus of elasticity in this case was (1.2%) compared to the base metal, and the greatest results were an increase in the average yield stress (16.9%) and maximum tensile stress (11.1%).

6.2 The impact of cutting depth on the friction stir process' microstructure

The original mineral's microstructure is made up of elongated grains with an average grain length of (187) m and a length-to-thickness ratio of (2.97), together with a heterogeneous distribution of precipitate particles as seen in **Figures 6** and 7.

The friction stir tool shoulder's spinning and pressing forces on the plate surface, along with the heat created by plastic deformation, cause a surface stirred layer (modified surface) to form beneath the tool shoulder to a depth of roughly a micrometer. The cutting depths (25, 50, 100, 150, 200, and 400 m) and stir layer were

No.	Travel speed *(mm/min)*	Tool rotation speed *(rpm)*	Specimen No.	Average Yield strength 0.2% offset y, *MPa σ*	Average Tensile strength *(S_T,MP)*	Elongation (%)	Modulus of Elasticity *(E, Gpa)*
			Base Metal	245	309	16	69.5
1	32	800	1	275	327	15	68.42
			2	271	333	16	69.61
2		1000	3	273	331	11	68.77
			4	279	335	14	71.87
3		1250	5	272	331	10	70.03
			6	274	336	12	71.07
4		1600	7	269	328	8	70.54
			8	272	335	9	71.27
5	63	800	9	271	333	9	60.91
			10	275	337	9	71.86
6		1000	11	280	337	10	63.45
			12	278	332	12	66.58
7		1250	13	268	335	10	66.71
			14	275	338	12	68.6
8		1600	15	272	329	12	63.06
			16	277	331	10	65.73

No.	Travel speed (mm/min)	Tool rotation speed (rpm)	Specimen No.	Average Yield strength 0.2% offset y, MPa σ	Average Tensile strength (S_T,MP)	Elongation (%)	Modulus of Elasticity (E, Gpa)
9	**80**	**800**	**17**	**285**	**338**	**12**	69.83
			18	**288**	**341**	**13**	70.89
10		**1000**	**19**	**277**	**328**	**13**	63.57
			20	272	325	13	65.33
11		1250	21	265	316	16	54.09
			22	271	337	13	71.95
12		1600	23	261	318	12	51.68
			24	263	322	12	68.42

Table 5.
The tests of the friction stir process of aluminum alloy (T6–6061) samples are compared for various rotational and linear speeds in order to select the optimal model for one stage and a depth of (400 m).

produced in two stages during the primary trials of the current study at an 800 rpm rotational speed and 80 mm/min travel speed, respectively. Both the first and second stages involve one pass each. From the chosen 800 rpm rotational speed with an 80 mm/min travel speed, **Figures 8–19** show all surfaces stirred layers created.

The reason for this is that the grains in the center of the friction area are smooth, equal in size along their axes, and smaller than the grains of the base metal. This is because the process of re-smoothing the crystalline sizes of the particles deposited in this area as a result of the plastic deformation caused by the high rotational movement of the friction tool and the occurrence of recrystallization as a result of the heat generated has occurred. As a result of the support's head (Shoulder) rubbing on the metal frame. Through a single analysis of the crystal structure of the friction process, as shown in the images of the microscopic structure of the alloy, Studying the crystal structure of the friction process for a single time allows us to observe that, while cutting from the surface of the original metal, there is

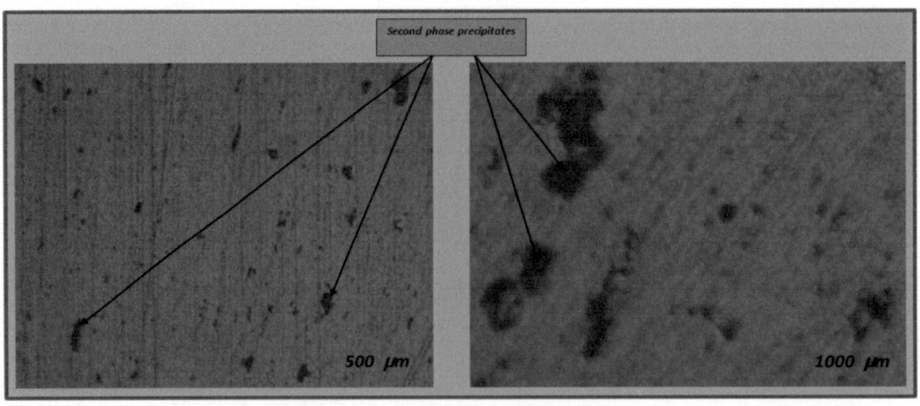

Figure 6.
Show different magnifications of the basic mineral's microstructure (AA6061-T6).

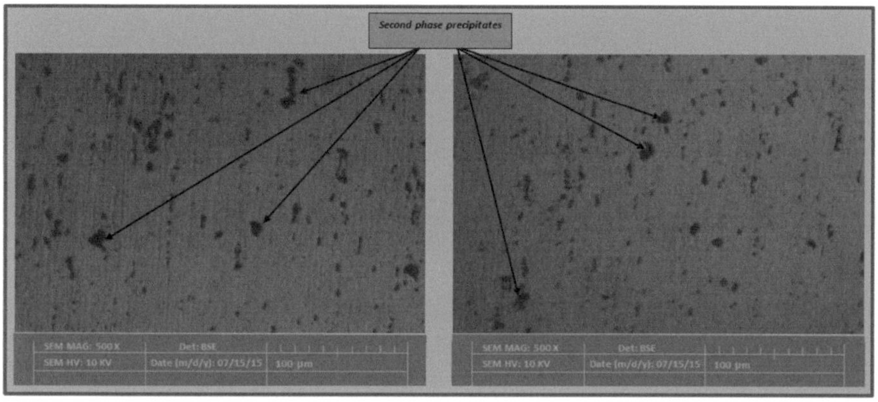

Figure 7.
Images captured by scanning electron microscopy of the aluminum alloy's base metal a backscattered electron micrograph and a secondary electron micrograph are also available.

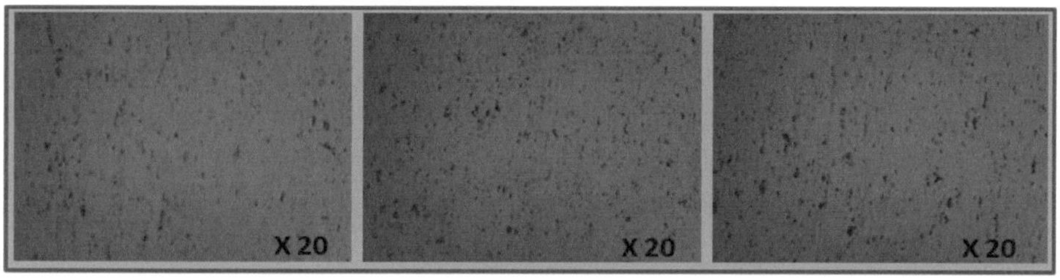

Figure 8.
Micrographs of the friction stir layer taken once at 0 µm depth, 800 rpm rotation speed, and 80 mm/min linear speed.

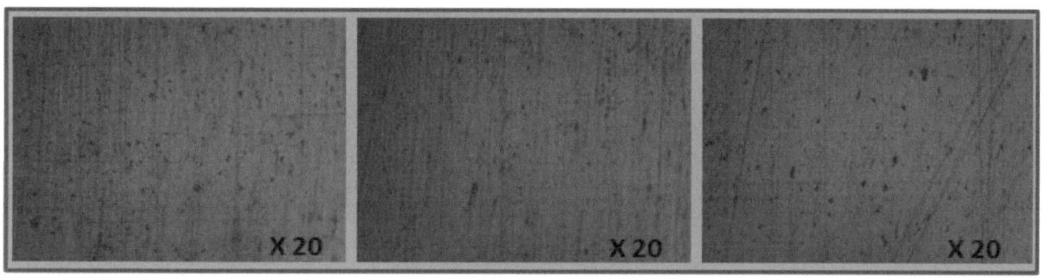

Figure 9.
Micrographs of the friction stir layer cut once at a depth of 25 µm at an 800 rpm rotational speed and an 80 mm/min linear speed.

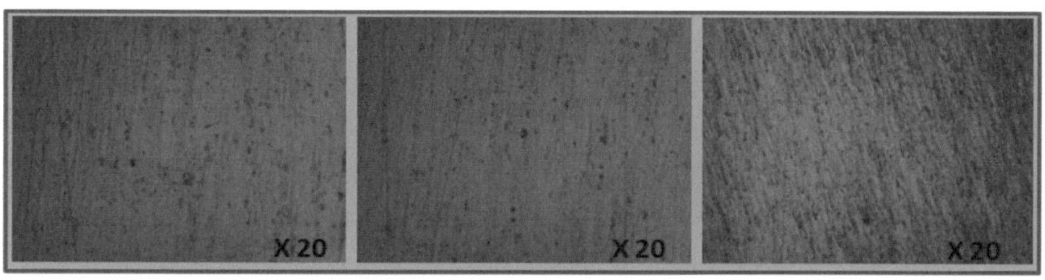

Figure 10.
Micrographs of the friction stir layer cut once at a depth of 50 µm at an 800 rpm rotational speed and an 80 mm/min linear speed.

Figure 11.
Micrographs of the friction stir layer cut once at a depth of 100 μm at an 800 rpm rotational speed and an 80 mm/ min linear speed.

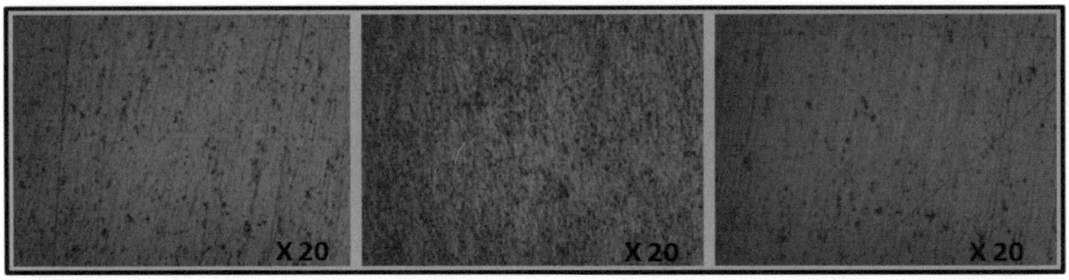

Figure 12.
Micrographs of the friction stir layer taken at a single 150 μm cutting depth, 800 rpm rotating speed, and 80 mm/ min linear speed.

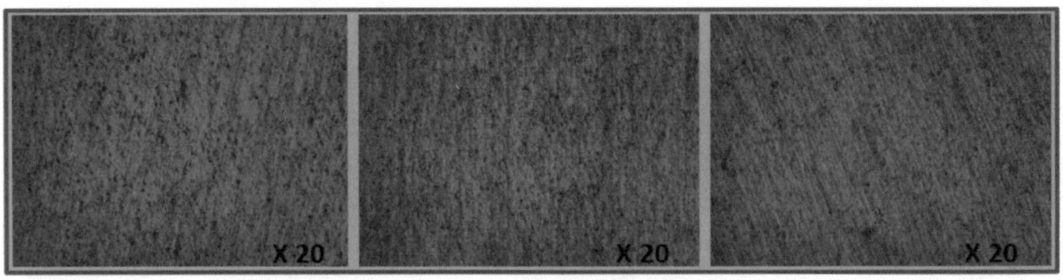

Figure 13.
Micrographs of the friction stir layer taken at a single 200 μm cutting depth, 800 rpm rotating speed, and 80 mm/ min linear speed.

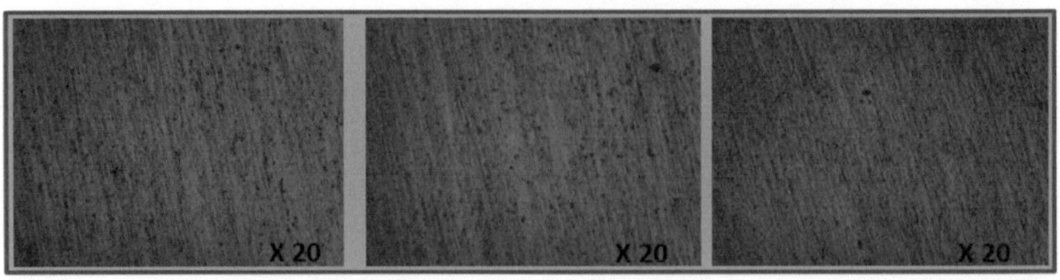

Figure 14.
Micrographs of the friction stir layer taken at a single 400 μm cutting depth, 800 rpm rotating speed, and 80 mm/ min linear speed.

Figure 15.
Micrographs of the friction stir layer taken twice at a cutting depth of 25 μm and at 80 mm/min in linear motion and 800 rpm in rotation.

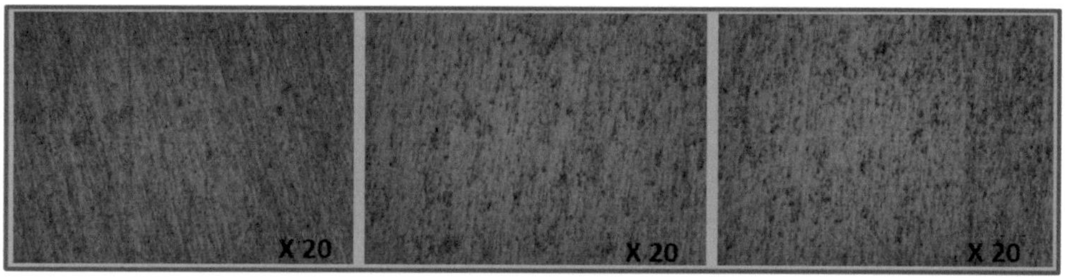

Figure 16.
Micrographs of the friction stir layer taken twice at a cutting depth of 50 μm and at 80 mm/min in linear motion and 800 rpm in rotation.

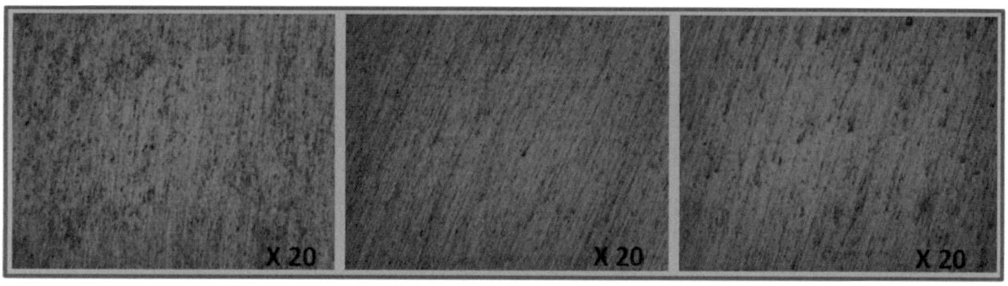

Figure 17.
Micrographs of the friction stir layer taken twice at a cutting depth of 100 μm and at 80 mm/min in linear motion and 800 rpm in rotation.

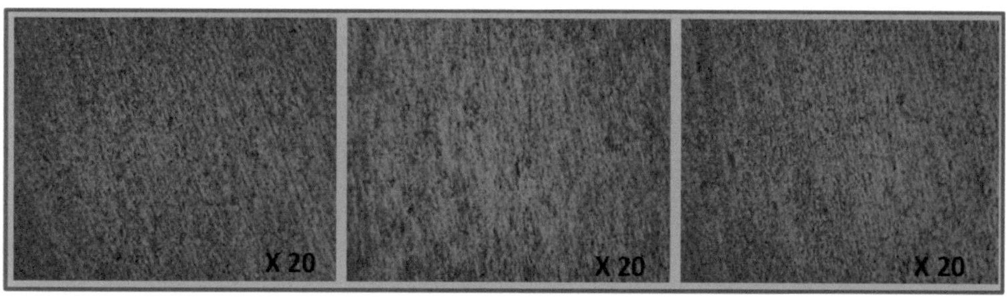

Figure 18.
Micrographs of the friction stir layer taken twice at a cutting depth of 200 μm and at 80 mm/min in linear motion and 800 rpm in rotation.

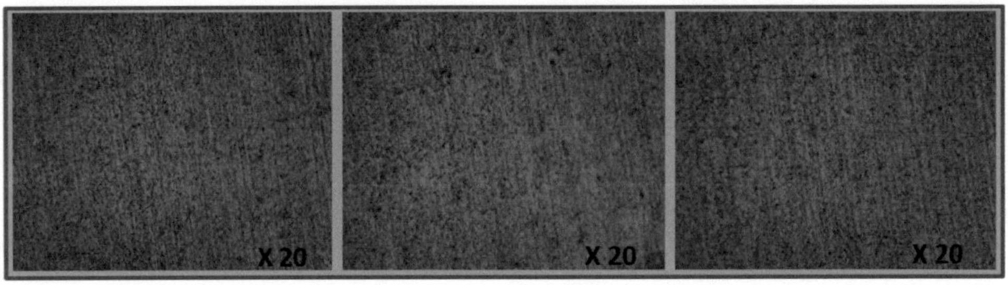

Figure 19.
Micrographs of the friction stir layer taken twice at a cutting depth of 400 μm and at 80 mm/min in linear motion and 800 rpm in rotation.

a percentage reduction in the size of the geometric flaws of the metal surface as follows:

When the depth of the cut was (25 m), the percentage of the size of the defects decreased (3.03% size), at (50 m, the percentage of decrease was approximately (3.32% size), at (100 m), the decrease was clearly greater, reaching (21.61% size), at (150 m), it was (30.42% size), at (200 m), it was (38.32%), and the highest decrease was recorded when the depth of the cut was (400 m).

The size of the fault grains in the metal surface decreased in comparison to the surface of the original metal in the following ways: The depth of cut at (25 m) was (4% size), the depth at (50 m) was (6% size), the depth at (100 m) was (10% size), the depth at (150 m) was (35% size), the depth at (200 m) was (38% size), and the depth at (400 m) had the largest percentage of grain size decrease, which amounted to (400% size).

Observe from the crystal structure and analysis that the geometric flaws have smaller grains and that the size of these grains reduces as the depth of cut increases. Furthermore, there is a greater proportion of shrinking and smallness in the friction process over two times than there is in the friction process over one time.

6.3 The impact of cutting depth on the friction stir process' mechanical characteristics

For the friction stir process of aluminum alloy samples (T6–6061), the necessary mechanical examination tests (hardness examination, tensile test, finding the value of the elastic modulus, and finding the value of yield strength) are carried out at a rotational speed (800 rpm) and a linear speed (mm/min80) for one and two times as well as for different depths of the process. Surface stir caused by friction. The outcomes are displayed in the **Tables 6** and 7.

The frictional mixing process's cutting depth and the hardness value are related in **Figure 20**. The figure clearly illustrates the relationship between the cutting depth and the hardness value in the case of the friction mixing process for two times, and it also indicates that this rise in hardness value remains considerable up to a depth of (300 mm). As the cutting depth is increased to stroke (300 mm), the hardness value increases dramatically, however this growth is drastically decreased. Additionally, a comparison between the single and double frictional mixing cases, where the hardness value rises as the cutting depth increases In addition, the amount of frictional mixing that occurs during a double frictional mixing process is greater than it is during a

Cutting depth *μm*	Specimen No.	Average Yield strength 0.2% offset y σ *MPa*	Average Tensile strength S_T *MPa*	Modulus of Elasticity E *GPa*	Elongation %	Hardness HV Kg/mm^2	Average Hardness HV Kg/mm^2
Base Metal		245	309	69.5	16	101.5	101.5
0	36	247	311	70	14.4	100.8	101.6
	37	259	310	71	14.6	102.4	
	38	260	313	67	15.1	101.7	
25	39	267	315	69	12.2	101.3	102.2
	40	263	318	67	9.7	101.1	
	41	272	316	69	8.0	104.2	
50	42	272	322	67	10.6	105.5	103.5
	43	270	325	68	11.5	101.6	
	44	269	322	69	15.1	103.4	
100	45	279	324	68	11.0	104.5	104.1
	46	280	327	69	11.9	103.1	
	47	277	331	69	12.9	104.7	
150	48	283	331	66	15.0	100.6	104.3
	49	282	326	70	10.8	108.1	
	50	278	336	69	10.2	104.2	
200	51	281	328	69	12.2	110.7	108.1
	52	286	339	70	12.6	108.1	
	53	282	334	67	17.4	105.6	
300	54	281	336	71	10.2	112.7	110.7
	55	285	330	67	9.1	111.8	
	56	288	341	71	8.7	107.6	
400	57	289	343	72	14.0	112.1	110.9
	58	282	333	69	13.1	108.8	
	59	284	339	70	13.6	111.7	

Table 6.
Compares the results of the friction stir experiments conducted on samples of aluminum alloy (T6–6061) at two different speeds: 800 rpm for rotation and 80 mm/min for linear motion, for one stage of the surface.

Cutting depth *μm*	Specimen No.	Yield strength 0.2% offset y σ MPa	Tensile strength S_T MPa	Modulus of Elasticity *E GPa*	Elong. %	Hardness HV Kg/mm^2	Average Hardness HV Kg/mm^2
Base Metal		245	309	68	16	101.5	101.5
50	60	275	324	62.33	11.9	105.9	103.5
	61	273	322	58	11.3	103.7	
	62	276	323	44	8.7	100.8	

Cutting depth *µm*	Specimen No.	Yield strength 0.2% offset y σ MPa	Tensile strength S_T MPa	Modulus of Elasticity *E GPa*	Elong. %	Hardness HV Kg/mm^2	Average Hardness HV Kg/mm^2
100	63	284	328	46.36	15.1	104.6	106.1
	64	279	330	44.77	13.8	109.6	
	65	285	331	49.78	13.2	104.1	
150	66	284	339	53.75	14.9	112.6	113.4
	67	287	340	59.42	14.4	115.6	
	68	284	337	54.77	15.6	112.1	
200	69	289	340	58.75	14.6	119.1	117.2
	70	284	342	58.64	14.8	115.2	
	71	286	340	59.33	12.2	117.3	
300	72	288	342	54.35	12.8	122.6	123.3
	73	292	344	52.15	14.6	124.1	
	74	289	340	53.47	13.7	123.1	
400	75	290	345	53.35	11.9	125.1	124.2
	76	288	344	50.5	11.3	120.5	
	77	295	345	52.48	8.7	127.1	

Table 7.
Compares the results of tests on the friction stir process for samples of aluminum alloy (T6–6061) at two different speeds: 800 rpm for rotation and 80 mm/min for linear motion.

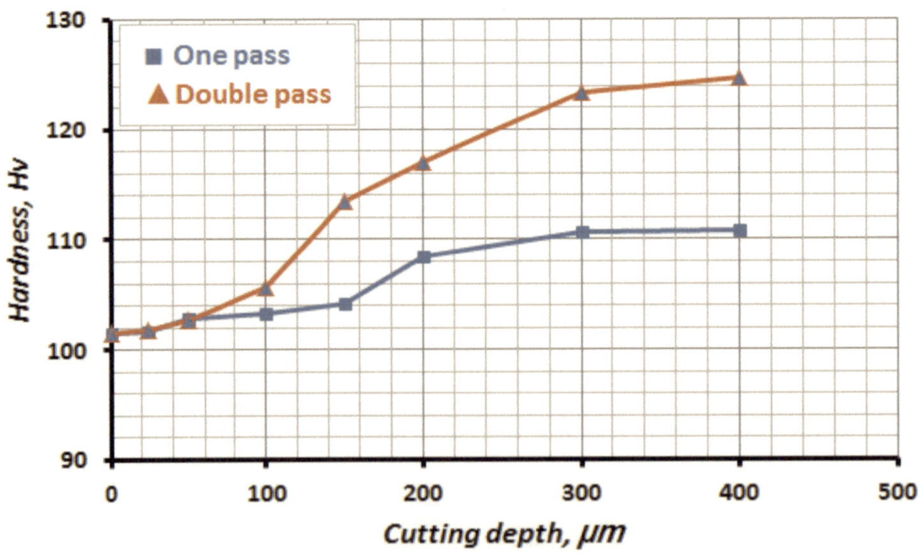

Figure 20.
Shows the connection between the hardness value and the frictional mixing process' cutting depth. Comparison of the mixing processes used once and twice.

single frictional mixing process. This is because the homogeneity of the crystals is greater, and the high temperature of the double frictional mixing process results in a reduction in engineering flaws and impurities.

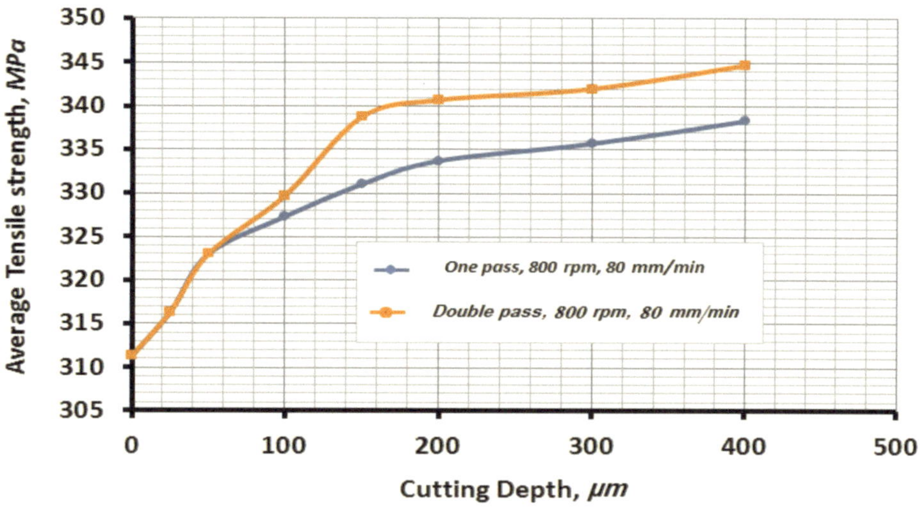

Figure 21.
Shows the relationship between the value of the tensile stress rate and the cutting depth of the friction mixing process. Comparison of the mixing processes used once and twice.

The maximum tensile stress rate increases with the increase in cutting depth, as shown in **Figure 21**, which contrasts the cases of the single and double friction mixing processes. Additionally, the double frictional mixing procedure is (2%) more efficient than the single frictional mixing technique.

7. Conclusions

The following conclusions can be reached after doing the study, reviewing, and discussing the findings:

1. The results of the examination and analysis show that the hardness increases with an increase in the cutting depth in the friction mixing processes; for one stage of frictional mixing, it increased by 9.3%, and for two stages, it increased by 22.4% compared to the base metal and at a depth of (400 m), and the increase is greater in the mixing process. With a rate of 13.1%, frictional mixing occurs twice from frictional mixing occurs once.

2. It is also clear from the microscopic structure of the samples that were subjected to the friction mixing processes that increasing the friction mixing process' cutting depth increased the homogeneity of the crystals and significantly reduced the number and size of engineering defects and impurities. Additionally, the reduction is greater when combining twice.

3. For one stage of the frictional mixing process, the yield strength increases by 18%, the tensile strength by 9.5%, and for two stages of the frictional mixing process, the yield strength increases by 20.4% and the tensile strength by 11.5%.

4. One stage of the friction mixing process improves the rate of hardness by 8.3% compared to the base metal, and two stages of the friction mixing process increase the rate of hardness by 18% compared to the base metal.

5. As the welding tool's rotating speed is increased, the bending resistance reduces while the welding table's linear speed increases. Using a friction mixing welding tool with a square indentation bump as opposed to one with a conical bump improves the bending resistance (cone).

Author details

Emad Toma Bane Karash* and Mohammad Takey Elias Kassim
Department of Mechanical Technologies, Mosul Technical Institute, Northern Technical University, Mosul, Iraq

*Address all correspondence to: emadbane2007@ntu.edu.iq

References

[1] Zakharov VV. Aluminum alloys: Some problems of the use of aluminum – Lithium alloys. Metal Science and Heat Treatment. 2003;**45**(1–2):49-54

[2] Akula DR. Characterization of Mechanical Properties and Study of Microstructure of Friction Stir Welded Joints Fabricated from Similar and Dissimilar Alloy of Aluminum. Columbia: Degree Master of Science, University of Missouri; 2007

[3] Odeshi AG, Adesola AO, Badmos AY. Failure of AA 6061 and 2099 aluminum alloys under dynamic shock loading. Engineering Failure Analysis. 2013;**35**: 302-314. DOI: 10.1016/j. engfailanal.2013.02.015

[4] Periyasamy YK, Perumal AV, Periyasamy BK. Optimization of process parameters on friction stir welding of AA7075-T651 and AA6061 joint using response surface methodology. Materials Research Express. 2019;**6**: 96558-96584

[5] Husain Mehdi RS. Mishra, Efect of friction stir processing on microstructure and mechanical properties of TIG welded joint of AA6061 and AA7075. Mater. Metallography Microstructure and Analysis. 2020;**9**(3):1-10. DOI: 10.1007/s13632-020-00640-7

[6] Refaai MRA, Reddy RM, Radha A, Christopher D. The influence of process parameters on the mechanical properties of friction stir-welded dissimilar Aluminium alloys AA2219 and AA7068. Advances in Materials Science and Engineering. 2022;**2022**, Article ID 3104199, 9 pages:1-9. DOI: 10.1155/2022/ 3104199

[7] Murali S, Chockalingam A, Suresh Kumar S, Remanan M. Production, characterization and friction stir processing of AA6063-T6/Al3Tip in-situ composites. International Journal of Mechanical and Production Engineering. 2018;**6**:399-406 View at: Google Scholar

[8] Manikandan R, Elatharasan G. Effect of process parameters on microstructural and mechanical properties of friction stir welded dissimilar aluminium alloys AA 6061 and AA 7075. International Journal of Rapid Manufacturing. 2020;**9**(1):1-15 View at: Publisher Site | Google Scholar

[9] Pasha A. Fabrication of surface metal matrix composite of AA7075 using friction stir processing. International Journal of Scientific Research in Science and Technology. 2022;**9**(3): 551-555 https://doi.org/10.32628/ IJSRST2293108

[10] Kumar K, Gulati P, Gupta A, Shukla DK. A review of friction stir processing of Aluminiumalloys using different types of reinforcements. International Journal of Mechanical Engineering and Technology. 2017;**8**(7): 1638-1651

[11] Salman JM. A comparative study an additives of nickel, cobalt, tin affecting the micros tructures and mechanical properties of Al-Zn-Mg-Cu alloys. Journal of Babylon University/ Engineering Sciences. 2014;**76**(3)218- 235. DOI: 10.13140/RG.2.2.12531.43043

[12] Murali VV, Satyanarayana K. Microstructure and mechanical properties of multipass friction stir processed aluminum silicon carbide metal matrix. International Journal of Scientific Engineering and Technology (ISSN: 2277-1581) Volume No. 4 Issue No. 2015;**2**:88-90

[13] Karthikeyan L, Senthilkumar VS, Balasubramanian V, Natarajan S. Mechanical property and microstructural changes during friction stir processing of cast aluminum 2285 alloy. Materials and Design. 2008;**30**: 2237-2242

[14] El-Danaf EA, El-Rayes MM, Soliman MS. Friction stir processing: An effective technique to refine grain structure and enhance ductility. Materials and Design. 2010;**31**:1231-1236

[15] ASTM E 384 – 99, Standard Test Method for Micro indentation Hardness of Materials. Conshohocken, PA, United States: ASTM; 2000. Standard_Test_Method_for_Microindentatio.pdf

[16] ASTM E8M – 04. Standard Test Methods for Tension Testing of Metallic Materials [Metric]. West Conshohocken, PA, United States: An American National Standard American Association State Highway and Transportation Officials Standard AASHTO No.: T68, ASTM International; 2004

[17] ASTM E 290 – 9. Standard Test Method for Bend Testing of Material for Ductility. USA, 2009. www.xycxie.com (chem17.com)

[18] ASTM E 23. Sub – Size Charp y, V Notched Specimens. Standard Test Methods for Notched Bar Impact Testing of Metallic Materials. USA, 2005

[19] Wang X, Lados D. Understanding the material flow mechanisms microstructure evolution defect formation relationships and effects on mechanical properties in friction stir welding of dissimilar aluminum alloys. Metallurgical and Materials Transactions A Published by ASM International. 2022; **54**:5. DOI: 10.1007/s11661-022-06921-w

[20] ASTM E23-18. Notched Bar Impact Testing of Metallic Materials1|PDF|Fracture|Temperature (scribd.com)

Chapter 10

Electrochemical Corrosion Study of Cold-Rolled AA8015-Alloy Processed by Reversing Cold Rolling Mill at Varying Surface Roughness

Olayinka Olaogun, Esther Titilayo Akinlabi and Cynthia Samuel Abima

Abstract

AA8015-alloy is a general-purpose aluminium alloy having a wide range of applications. Electrochemical corrosion testing of cold-rolled AA8015-alloy processed by reversing cold-rolling mill at different surface roughness in natural seawater were investigated. The AA8015-alloy utilised in this study was cold-rolled in a reversible Achenbach cold-rolling mill in four number of passes to a gauge thickness of 1.2 mm. This industrial cold rolling process was achieved at Tower Aluminium Rolling Mill, Sango-Ota, Nigeria. The different surface roughness's each with three cold mounted samples, 1.54 μm, 0.83 μm, 0.18 μm and 0.04 μm were achieved on automated polishing machine using 320-grit, 800-grit, 1200-grit and diamond abrasive MD-Mol respectively. Electrochemical corrosion experiments were conducted on the samples in natural seawater using a computer-controlled potentiostat in an open polarisation cell set-up at room-temperature. The corrosion behaviour on surface morphologies of the samples was observed by high-mega-pixel camera and scanning electron microscope. Findings reveal asymmetric polarisation curves, and the polarisation resistance increases as the surface roughness decreases. Consequently, corrosion-rate reduces as the surface get smoother and EDS elemental analysis shows the existence of insoluble sulphate and chloride complexes formed on the surfaces. Conclusively, surface roughness affects the corrosion resistance of cold-rolled AA8015-alloy in natural seawater.

Keywords: cold-rolled AA8015-alloy, electrochemical corrosion, reversible rolling mill, surface roughness, corrosion rate analysis

1. Introduction

The useful lifespan of any component or device is always established at its design and manufacturing stage [1]. In extractive metallurgy, large billets have to be reduced by mechanical deformation processes such as forging, rolling and extrusion for further reduction and change in their shapes. There are three basic temperature ranges in metal forming at which the metal (workpiece) can be formed which are hot working, warm working and cold working [2]. Cold working is a strengthening mechanism that involves plastic deformation. This strengthening mechanism is mostly utilised in ductile metals. In addition, cold working takes place when the processing temperature of the mechanical deformation of the ductile metal is below the recrystallization temperature [3–5]. Cold working in mechanical rolling process is eminent as compared to pressing, drawing, spinning and extruding.

The cold working in rolling process, known as cold rolling, involves deforming the ductile metal by using rolls at low temperatures, especially at temperatures below the recrystallization temperature of the specific metal. Cold rolling processes are achieved by using rolling mills to produce metal sheets of a certain required thickness. The vast majority of cold rolled metal is in the form of flat rolling. Worth noting is the fact that cold rolling has enormous benefits in its ability to manufacture products from relatively large pieces of metal at very high speed in a continuous manner with good surface finish, highly accurate tolerances and stronger products. In addition, cold rolling process eliminates shrinkage effect, increases hardness and elastic limit. It also promote decrease in ductility due to strain hardening effect [2, 4, 6, 7].

There has been advances in research on corrosion behaviour of cold rolled metal-alloys in various solutions reported. Liu et al. [8] carried out electrochemical measuring technique on metastable Cr-Mn-Ni-N austentic stainless steel in acidic medium. Their findings shows that Fe and Cr dissolutes on the stainless steel surface in the process of corrosion which is clearly facilitated by the cold rolling deformation. Studies by Ma et al. [9] demonstrated that high strained cold rolling deformation such as 70 and 90 percent reduction in thickness in Ta-4 W alloy improves corrosion resistance due to preferential crystallographic orientations. More studies [10–13], also confirms that cold rolling deformation effect in metal-alloys greatly influence the corrosion current density and corrosion potentials.

Moreover, open literature reports evidence of aluminium and its alloys corroding under typical applications. To mention a few. Rao et al. [14] in their literature review on stress-corrosion cracking, confirm that 2xxx, 5xxx and 7xxx aluminium alloys are susceptible to stress-corrosion cracking. Moreover, a review on the corrosion inhibition performance evaluation for aluminium and its alloys in chloride and alkaline solutions by Xhanari and Finsgar [15], confirms evidence of corrosion in aluminium. Further scholarly articles [16–25] reported also indicate that aluminium and its alloys corrode, despite its formation of a natural oxide layer that is chemically inert. However, corrosion study on aluminium 8-series especially aluminium 8015-alloy is yet to be reported.

Conducting effect of corrosion on varying degree of surface roughness in cold rolled metal reduction has been a grey area of research. Therefore, this research work focus on an extensive in-depth experimental investigation on the corrosion behaviour of cold-rolled AA8015-alloy in natural seawater solution at varying surface roughness condition.

2. Achenbach reversing mill and cold rolling process

The Achenbach reversing cold rolling mill utilised in this study is a 4-high single stand unit at Tower Aluminium Rolling Mill, Sango-Ota, Nigeria. The 4-high reversing mill detail description and pictorial representation are presented in **Table 1** and **Figure 1** respectively.

The annealed coiled AA8015-alloy with 7 mm sheet thickness is heated up to approximately 120°C before been loaded into the deforming rolls at industrial ambient temperature. The alloy sheet was cold rolled successively at thermal equilibrium maintained at about 70°C by passing it back and forth in four successive pass

	Description
Type	Single Stand Reversing 4Hi Cold-Rolling Mill
Make	Achenbach
Strip width	1250 mm
Speed	485 m / minute
Inner coil diameter	600 mm
Outer coil diameter	1520 mm
Entry thickness	Max. 8 mm
Exit thickness	Min. 0.20 mm
Hydraulic System Pressure	40 Bar
Load Capacity	12,000 kN

Table 1.
Description of Achenbach 4-high reversing cold rolling mill at towers aluminium rolling mill.

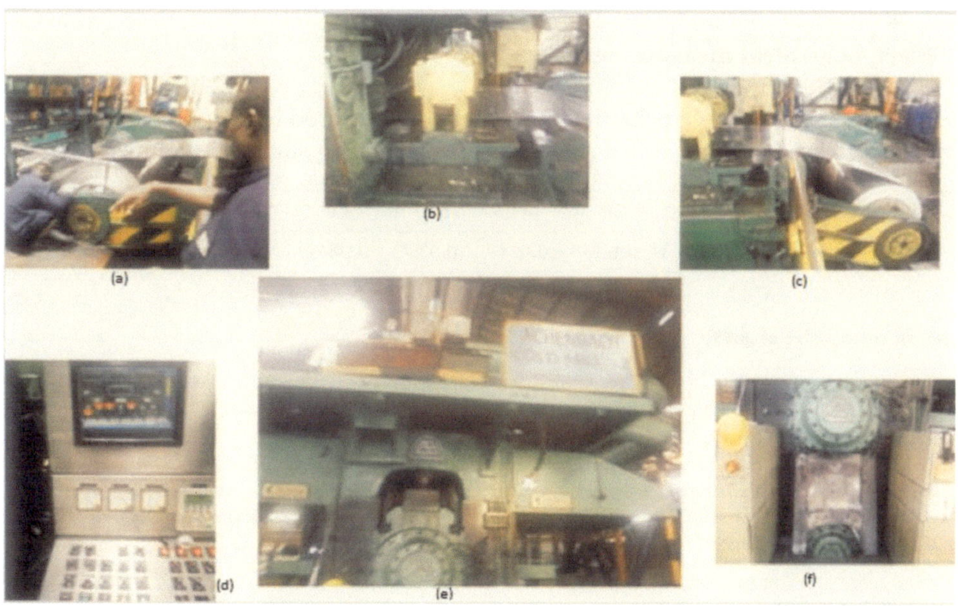

Figure 1.
Pictures of Achenbach 4-high reversible cold rolling mill at tower aluminium rolling mill; (a) left-end of the loaded coil, (b) cold rolling process, (c) right-end of the loaded coil, (d) computer numeric control unit, (e) bigger view of the 4-high reversible cold mill, (f) end view of the rolls.

schedules in alternate directions until desired thickness of 1.2 mm is attained. The pass schedules were chosen to maintain reasonable constant drive power and rolling force during successive passes.

3. Experimental details

3.1 Material, sample preparation and electrolyte solution

The cold-rolled AA8015-alloy specimen was investigated using Bruker Elemental Optical Spectrometry and the obtained elemental composition is given in **Table 2**. The cold-rolled alloy specimen was square-cut using automated Mecatome T300 cutting machine embedded with a 10S25 cut-off wheel based on the inscribed markings on the surface having dimensions 10 mm by 10 mm. Twelve corrosion specimen samples were prepared according to ASTM standard G1–03 [26] and cold mounted using EpoFix Resin and hardener following Struers application note for cold mounting procedures [27]. Further surface treatment was done on the cold mounted samples using SiC emery paper of varying 320, 800 and 1200 grit and diamond polished with MD-Mol disc surface on automated grinding/polishing machine. Four different surface conditions were attained after cleaning with distilled water and acetone. The surface roughness (R_a) was determined with HOMMEL-ETAMIC TURBO roughness and contour metrology, for each surface conditions given in **Table 3**.

A natural seawater electrolyte solution was utilised in the experiment with a pH of 7.04, taken from the Sea at Durban, South Africa. **Table 4** depicts the chemical composition of the seawater at 3.5% salinity. The major ion in parts per million (ppm) is shown.

3.2 Electrochemical measurements

Potentiodynamic electrochemical technique was utilised in determining the corrosion behaviour of the cold-rolled AA8015-alloy at varying surface roughness following

Element	Si	Fe	Cu	Mn	Mg	Cr	Zn	Ti	Al
Weight %	0.478	1.338	0.146	0.074	0.035	0.0067	0.072	0.011	balance

Table 2.
Chemical composition of AA8015-alloy in weight per cent.

Cold-rolled AA8015 alloy	Surface roughness value with 320 grit SiC paper (R_a)	Surface roughness value with 800 grit SiC paper (R_a)	Surface roughness value with 1200 grit SiC paper (R_a)	Surface roughness value with diamond polishing (R_a)	Units
Cold-rolled samples	1.54	0.83	0.18	0.04	μm

Table 3.
Surface roughness values of cold-rolled AA8015-alloy corrosion samples under different surface conditions.

Element	PPM in sea water
Chloride (Cl^-)	18,980
Sodium (Na^+)	10,556
Sulphate (SO_4^{2-})	2649
Magnesium (Mg^{2+})	1262
Calcium (Ca^{2+})	400
Potassium (K^+)	380
Bicarbonates (HCO_3^-)	140
Strontium (Sr^{2+})	13
Bromide (Br^-)	65
Borate (BO_3^{3-})	26
Fluoride (F^-)	1
Silicate (SiO_3^{2-})	1
Others	—

Table 4.
Major ion composition of seawater at 3.5% salinity [28].

ASTM standard G5 [29]. The potentiodynamic polarisation curves were generated using the Ivium Compact-Stat Potentiostat computer-controlled with accustomised Ivium corrosion analysis software to produce Tafel fit lines. Electrochemical experiments were conducted for three samples of the cold-rolled alloy at each surface roughness carried out at room temperature in an open glass cell containing 200 ml solution of natural seawater. This is important to confirm reproducibility and precision. Open circuit potential (E_{oc}) measurement of each sample of the cold-rolled

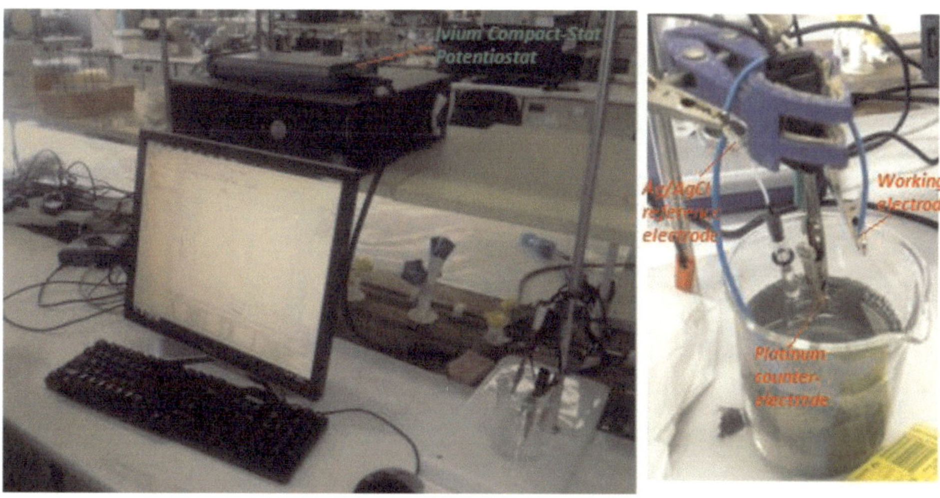

(a) (b)

Figure 2.
Electrochemical corrosion experimental set-up; (a) Ivium compact-stat Potentiostat, (b) an open glass cell system.

aluminium 8015-alloy at varying surface roughness was first determined, according to ASTM standard G69 [30]. The polarisation cell set-up for determination of E_{oc} consist of connections of the working electrode (prepared cold-rolled aluminium 8015-alloy samples at different surface roughness) and the silver/silver chloride reference electrode in natural seawater electrolyte solution to the potentiostat. Subsequent potentiodynamic experiments carried out utilised a conventional three-electrode polarisation cell that includes a platinum counter-electrode. A linear potential sweep in the anodic direction was performed at a scan rate of 0.167 mV/s, starting from 250 mV below the E_{oc} and terminating at 250 mV above the E_{oc}. The scanning electron microscope images were recorded to ascertain the interaction of seawater medium with the alloy surface using TESCAN VEGA Scanning Electron Microscope with Energy Dispersive X-ray Spectroscopy (EDS). **Figure 2** shows the experimental polarisation cell set-up accordingly.

4. Results and discussion

4.1 Open circuit potential (E_{oc}) and potentiodynamic polarisation result

The measured potential of the working electrode (cold-rolled AA8015-alloy) at varying surface roughness is given in **Figure 3**, as a function of time for an hour duration. The variations of the open circuit potential against Ag/AgCl with the time plot for the three samples show a steady state variation almost throughout the time duration with relatively stable drifting within 0.1 V for samples with surface roughness 1.54 μm and 0.83 μm, and 0.05 V for samples with surface roughness 0.18 μm and 0.04 μm respectively. The open circuit potential (E_{oc}) values for the three samples at each surface roughness all show electronegative potentials, given in **Table 5**. These steady-state E_{oc} values were taken at the last 3600 seconds. The anodic and cathodic reactions on the alloy surface are in equilibrium at these potentials.

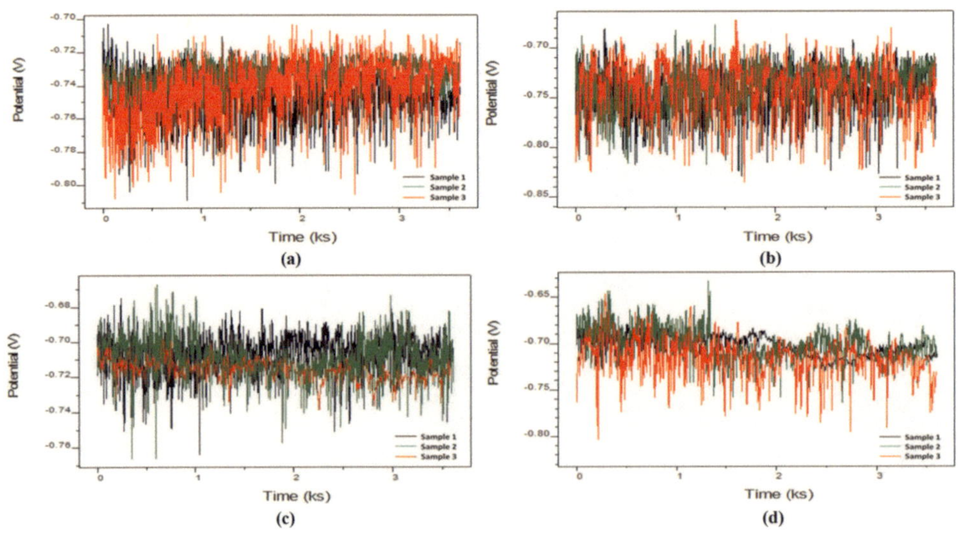

Figure 3.
Variation of open circuit potential with time for cold-rolled aluminium 8015-alloy at (a) 1.54 μm; (b) 0.83 μm; (c) 0.18 μm and; (d) 0.04 μm surface roughness's in natural seawater.

Cold-rolled AA8015 alloy samples	Surface roughness Ra ≈ 1.54 μm	Surface roughness Ra ≈ 0.83 μm	Surface roughness Ra ≈ 0.18 μm	Surface roughness Ra ≈ 0.04 μm	Units
Sample 1	−0.739	−0.740	−0.713	−0.734	V
Sample 2	−0.714	−0.728	−0.723	−0.710	V
Sample 3	−0.740	−0.759	−0.703	−0.714	V

Table 5.
Open circuit potentials (E_{oc}) for cold rolled AA8015 with varying surface roughness.

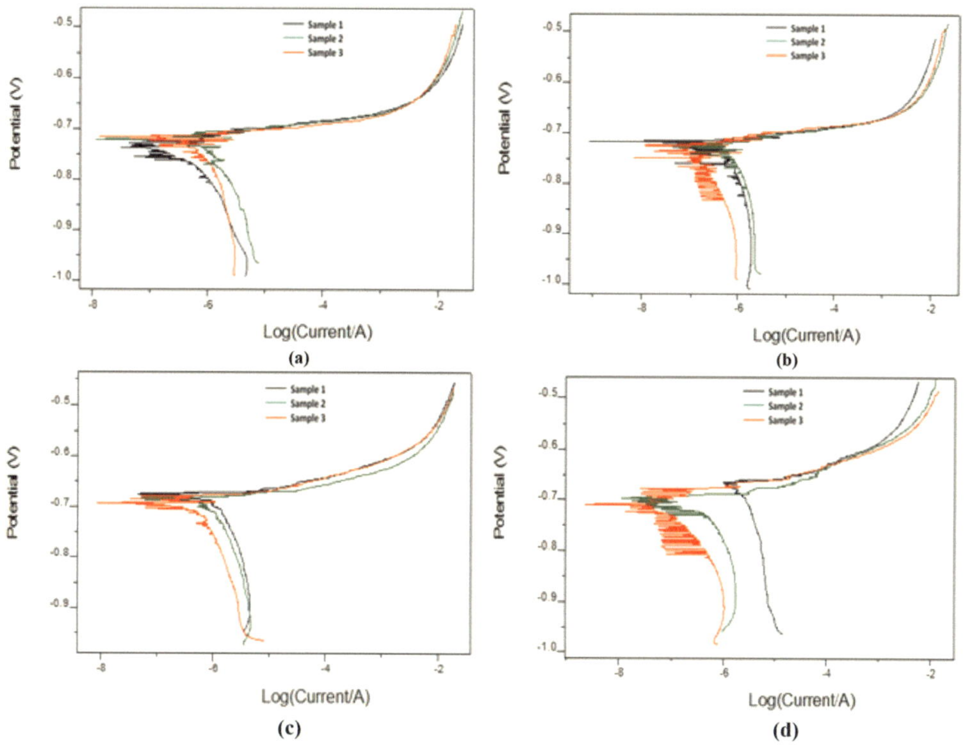

Figure 4.
Polarisation curves of cold-rolled AA8015-alloy immersed in natural sea water at surface roughness's (a) Ra ≈ 1.54 μm; (b) Ra ≈ 0.83 μm; (c) Ra ≈ 0.18 μm; (d) Ra ≈ 0.04 μm.

Likewise, **Figure 4** shows the polarisation curves generated for each potentiodynamic polarisation test for all the samples. Observations for each curve shows similarity in shape and are asymmetric. Furthermore, the Tafel behaviour on the cathodic side extends over a wider potential range than on the anodic side. In addition, the anodic current rose more steeply with changes in potential than the cathodic current. This indicates that reduction reaction occurs at slower rate than oxidation reaction. Therefore, cathodic reaction controls the electrochemical corrosion of the cold-rolled AA8015-alloy. The drifting occurrence in the polarisation curves confirms electrochemical noise. These suggest indication of sudden inert oxide-layer film rupture causing dissolution of the AA8015-alloy.

4.2 Corrosion rate analysis

The Tafel plot analysis on the cold-rolled AA8015-alloy samples at each surface roughness condition are presented in **Figures 5–8**. The current-potential data obtained

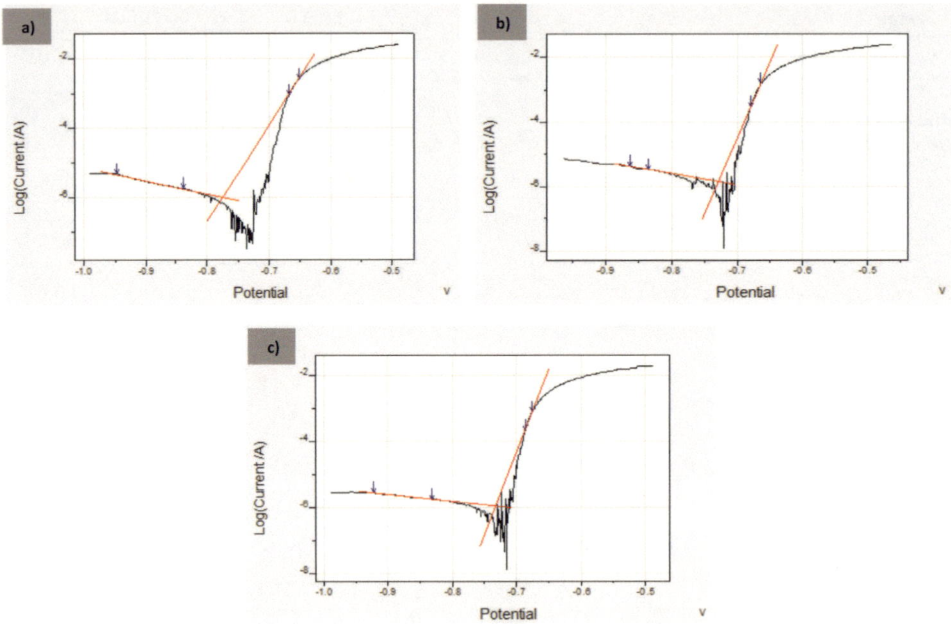

Figure 5.
Tafel analysis of cold-rolled aluminium 8015-alloy at $R_a \approx 1.54$ *μm. (a) Sample 1; (b) sample 2; (c) sample 3.*

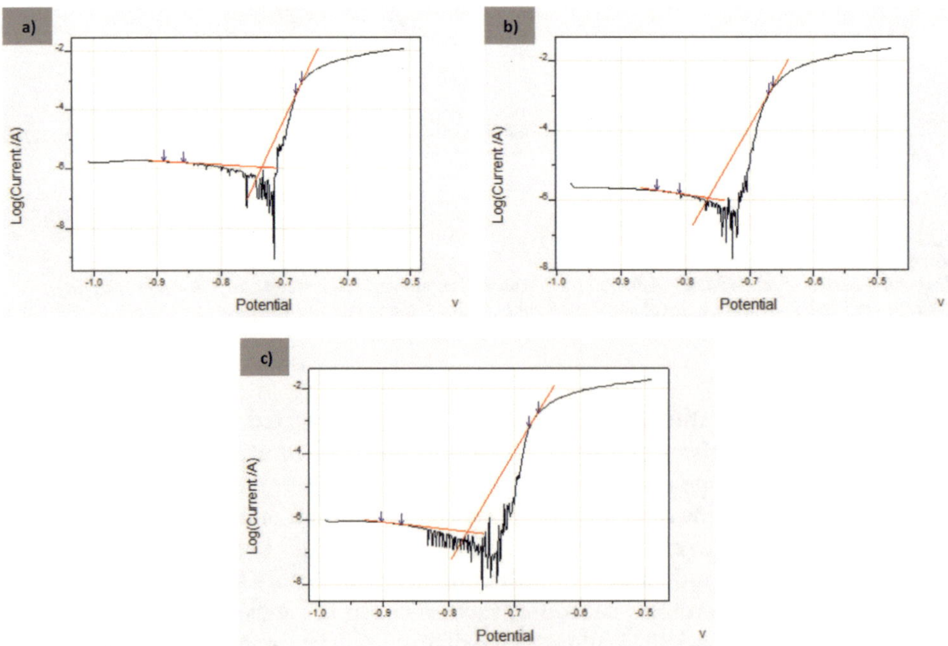

Figure 6.
Tafel analysis of cold-rolled aluminium 8015-alloy at $R_a \approx 0.83$ *μm. (a) Sample 1; (b) sample 2; (c) sample 3.*

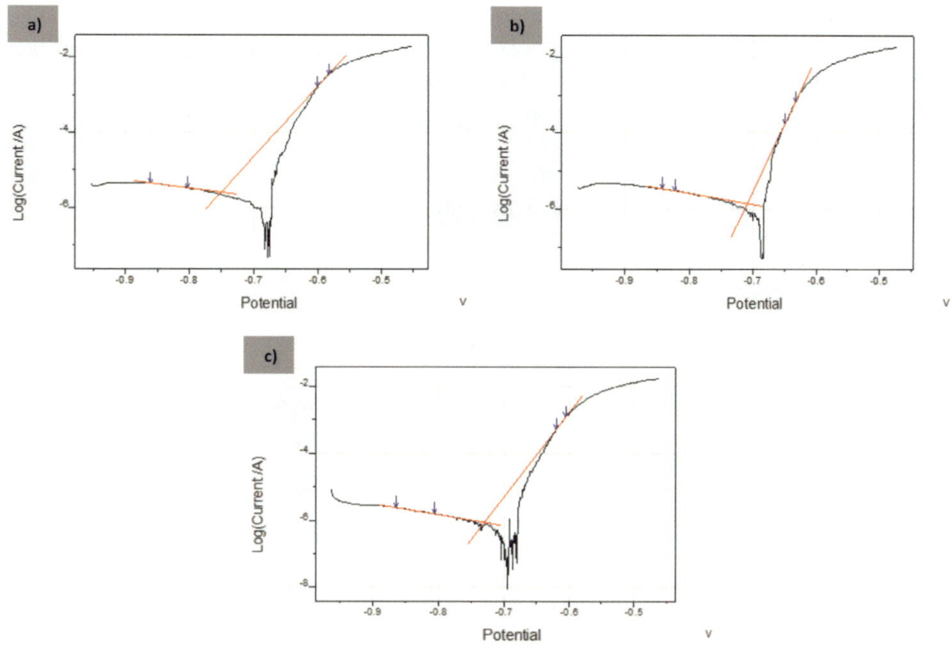

Figure 7.
Tafel analysis of cold-rolled aluminium 8015-alloy at $R_a \approx 0.18$ *μm. (a) Sample 1; (b) sample 2; (c) sample 3.*

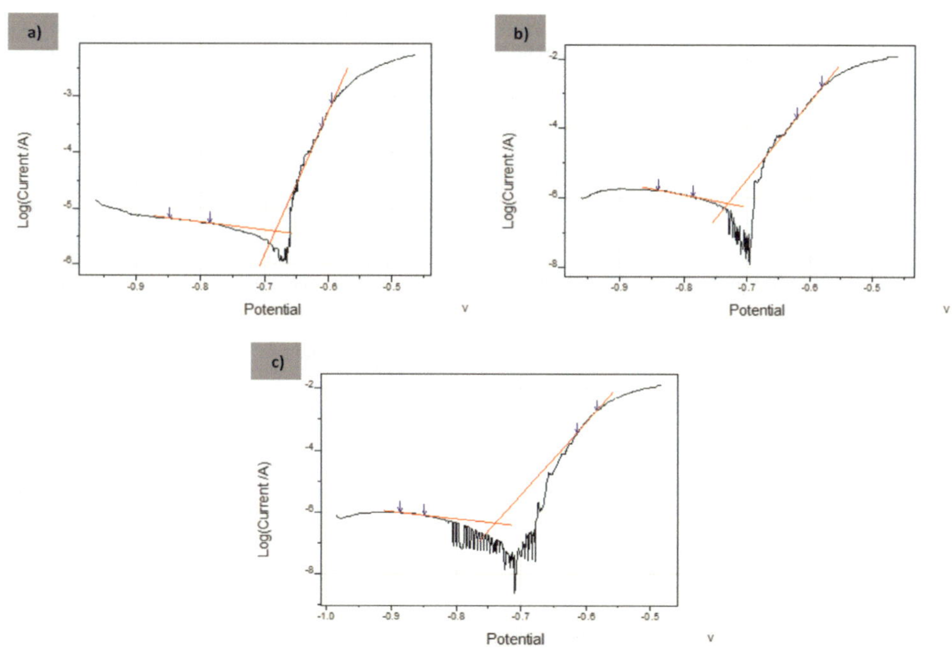

Figure 8.
Tafel analysis of cold-rolled aluminium 8015-alloy at $R_a \approx 0.04$ *μm. (a) Sample 1; (b) sample 2; (c) sample 3.*

were plotted as logarithms of current against potential for both the anodic and cathodic branches. Straight lines that best fit the data at high potentials were achieved at selected potential range markers on both anodic and cathodic curves. The point of intersection yields the corrosion current density (I_{corr}) and the corrosion potential (E_{corr}). The corresponding electrochemical kinetic parameters and corrosion rate for each surface roughness condition of the cold-rolled AA8015-alloy in natural seawater solution were computed and given in **Tables 6–9**.

Corrosion parameters	Sample 1	Sample 2	Sample 3
E. corr (V)	−0.7746	−0.7289	−0.7324
i cor. (A)	9.78E-07	1.42E-06	1.16E-06
I cor. (A/cm^2)	9.78E-07	1.42E-06	1.16E-06
Rp (Ohm)	14,040	6051	7173
ba (V/dec)	0.036	0.021	0.02
bc (V/dec)	0.266	0.288	0.485
C. Rate (mm/y)	**0.01063**	**0.01538**	**0.01266**

Table 6.
Corrosion rate analysis of cold-rolled AA8015-alloy at $R_a \approx 1.54\ \mu m$.

Corrosion parameters	Sample 1	Sample 2	Sample 3
E. corr (V)	−0.735	−0.7639	−0.7708
i cor. (A)	1.17E-06	1.19E-06	4.23E-07
I cor. (A/cm^2)	1.17E-06	1.19E-06	4.23E-07
Rp (Ohm)	8021	10,650	28,580
ba (V/dec)	0.022	0.032	0.03
bc (V/dec)	0.813	0.379	0.443
C. Rate (mm/y)	**0.01268**	**0.01297**	**0.004599***

Table 7.
Corrosion rate analysis of cold-rolled AA8015-alloy at $R_a \approx 0.83\ \mu m$.

Corrosion parameters	Sample 1	Sample 2	Sample 3
E. corr (V)	−0.7509	−0.7086	−0.729
i cor. (A)	2.53E-06	1.39E-06	8.81E-07
I cor. (A/cm^2)	2.53E-06	1.39E-06	8.81E-07
Rp (Ohm)	8236	8130	17,130
ba (V/dec)	0.054	0.028	0.039
bc (V/dec)	0.458	0.341	0.309
C. Rate (mm/y)	**0.02749***	**0.01516**	**0.009581**

Table 8.
Corrosion rate analysis of cold-rolled AA8015-alloy at $R_a \approx 0.18\ \mu m$.

Corrosion parameters	Sample 1	Sample 2	Sample 3
E. corr (V)	−0.683	−0.731	−0.7398
i cor. (A)	3.89E-06	7.11E-07	4.38E-07
I cor. (A/cm^2)	3.89E-06	7.11E-07	4.38E-07
Rp (Ohm)	4089	23,520	38,160
ba (V/dec)	0.039	0.044	0.043
bc (V/dec)	0.721	0.286	0.407
C. Rate (mm/y)	**0.04223***	**0.00773**	**0.004764**

Table 9.
Corrosion rate analysis of cold-rolled AA8015-alloy at $R_a \approx 0.04$ *μm.*

Surface roughness, R_a (μm)	Polarisation resistance, R_p (kΩ)	Corrosion rate, (mm/yr)
1.54	9.088	0.01289
0.83	9.336	0.012825
0.18	12.630	0.0123705
0.04	30.840	0.006247

Table 10.
Mean values of polarisation resistance and corrosion rate of cold-rolled AA8015-alloy at different surface roughness values.

The mean value calculated results in **Table 10**, reveals the effect of polarisation resistance (R_p) and corrosion rate on the surface roughness of cold-rolled AA8015-alloy in natural seawater solution at room temperature. Significant observation reveals increase in R_p values as the cold-rolled alloy surface roughness get smoother. Low R_p value of 9.088 kΩ recorded for surface roughness, $R_a \approx 1.54$ μm shows low resistance to corrosion attack as compared to high R_p value of 30.84 kΩ recorded for surface roughness, $R_a \approx 0.04$ μm. This is evidence in the visual inspection macrograph after electrochemical corrosion. See **Figure 9**. Similarly, the rate of corrosion decreases as the surface roughness reduces.

4.3 Visual inspection and SEM analysis

The macrographs in **Figure 9** were taken after electrochemical corrosion experiment. The presence of black spots on the surface reveals the extent of the attack in the form of localised corrosion. The magnitude of dissolution of the cold-rolled aluminium alloy in natural seawater solution revealed in the macrographs confirms increase in corrosion resistance as the surface roughness becomes smoother.

Further microstructural analysis using Scanning Electron Microscope (SEM) revealed corrosion by pitting in all the surface roughness conditions, given in **Figures 10–13**. In addition, substantial insoluble substrate complexes were observed, confirming evidence of corrosion. However, SEM images at surface roughness, $R_a \approx 1.54$ μm is without the presence of insoluble substrate complex. This could be due to the high mechanical surface flaws because of the 320-grit SiC paper. Moreover, EDS

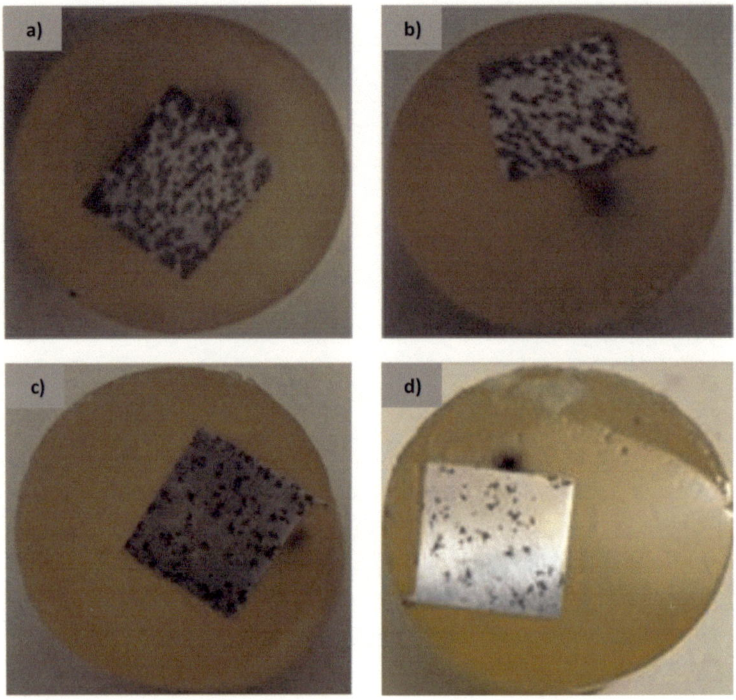

Figure 9.
Macrographs showing localised corrosion attack on cold-rolled AA8015-alloy surface samples after electrochemical corrosion. (a) $R_a \approx 1.54\ \mu m$; *(b)* $R_a \approx 0.83\ \mu m$; *(c)* $R_a \approx 0.18\ \mu m$ *(d)* $R_a \approx 0.04\ \mu m$.

analysis in **Figure 14** shows sulphur and chlorine ions present in the insoluble substrate confirming the adsorption chloride and sulphur molecules in seawater at the defective spots were oxide layer is dissolved.

5. Conclusions and future work

Electrochemical corrosion of the cold-rolled AA8015-alloy at varying surface roughness in natural seawater was investigated. This was shown in the open circuit potential and potentiodynamic polarisation. Outcome revealed:

1. Evidence of pitting corrosion

2. Subsequent surface roughness conditions examined shows that surface roughness affects the corrosion resistance of cold-rolled AA8015-alloy

3. Presence of insoluble substrate on AA8015-alloy surface in corrosion process

4. Corrosion resistance increases with high percent reduction of cold rolled AA8015-alloy

For in-depth analysis and to understand the active corrosion characteristics of cold rolled AA8015-alloy and in addition to the relationship between the microstructural

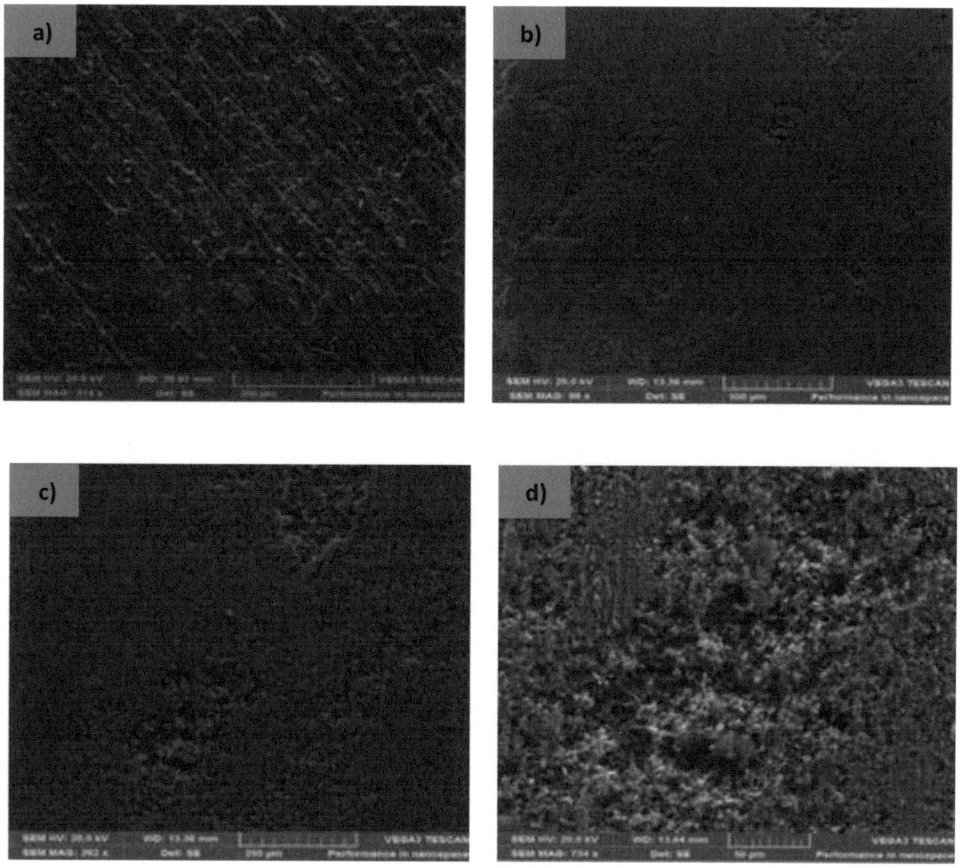

Figure 10.
SEM images of corroded cold-rolled AA8015 sample at $R_a \approx 1.54$ *µm. (a) Image before corrosion at 314-x magnification; (b-d) images after corrosion at increased magnifications.*

evolution and corrosion behaviour of cold rolled AA8015 in natural sea water. Future investigation using X-ray diffraction (XRD), electron back scatter diffraction (EBSD) and X-ray photoelectron spectroscopy (XPS) are recommended.

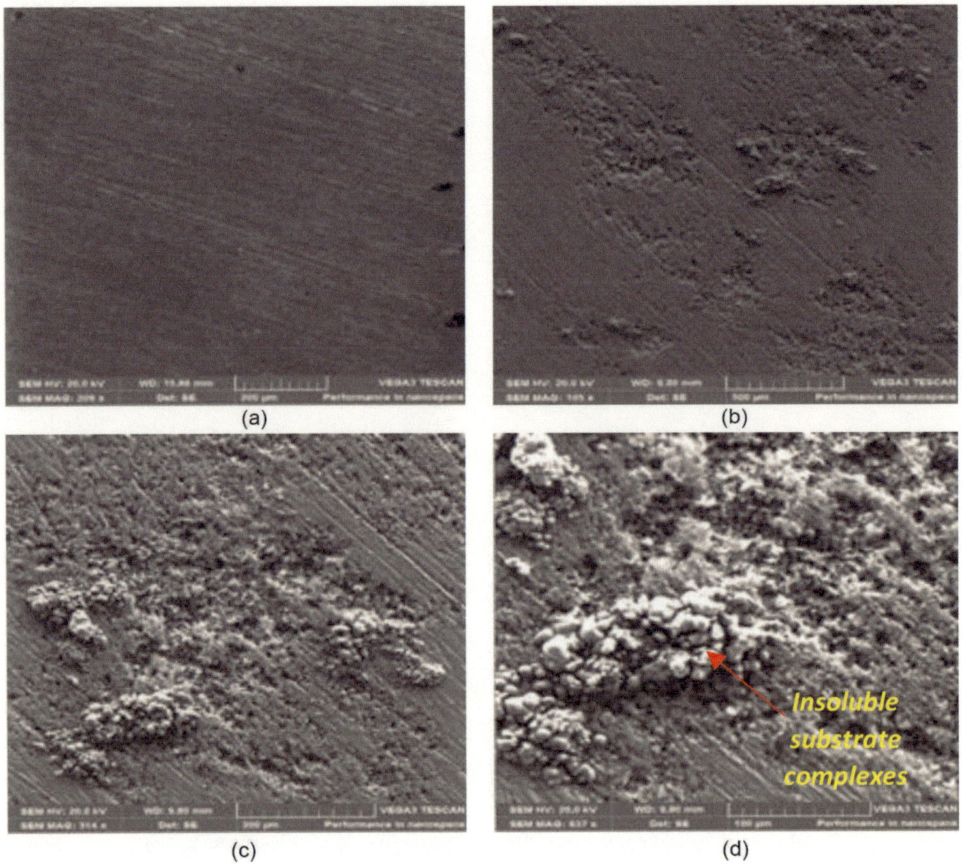

Figure 11.
SEM images of corroded cold-rolled AA8015 sample at $R_a \approx 0.83$ µm. (a) Image before corrosion at 209-x magnification; (b-d) images after corrosion at increased magnifications.

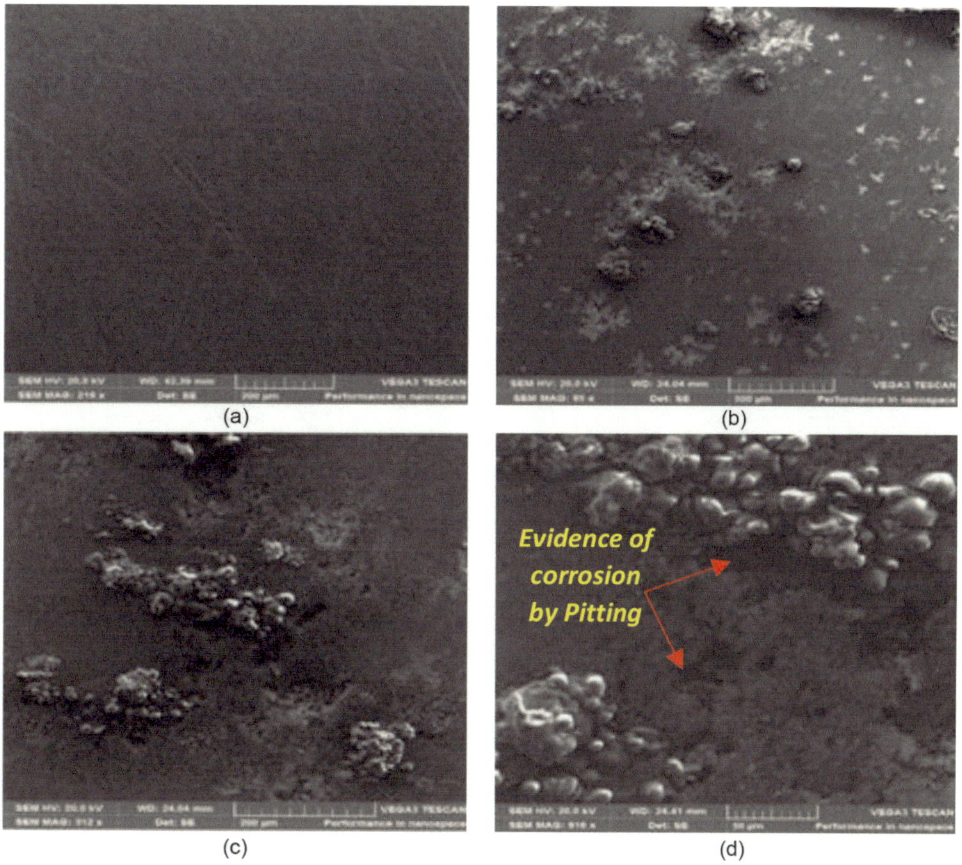

Figure 12.
SEM images of corroded cold-rolled AA8015 sample at $R_a \approx 0.18$ *μm. (a) Image before corrosion at 219-x magnification; (b-d) images after corrosion at increased magnifications.*

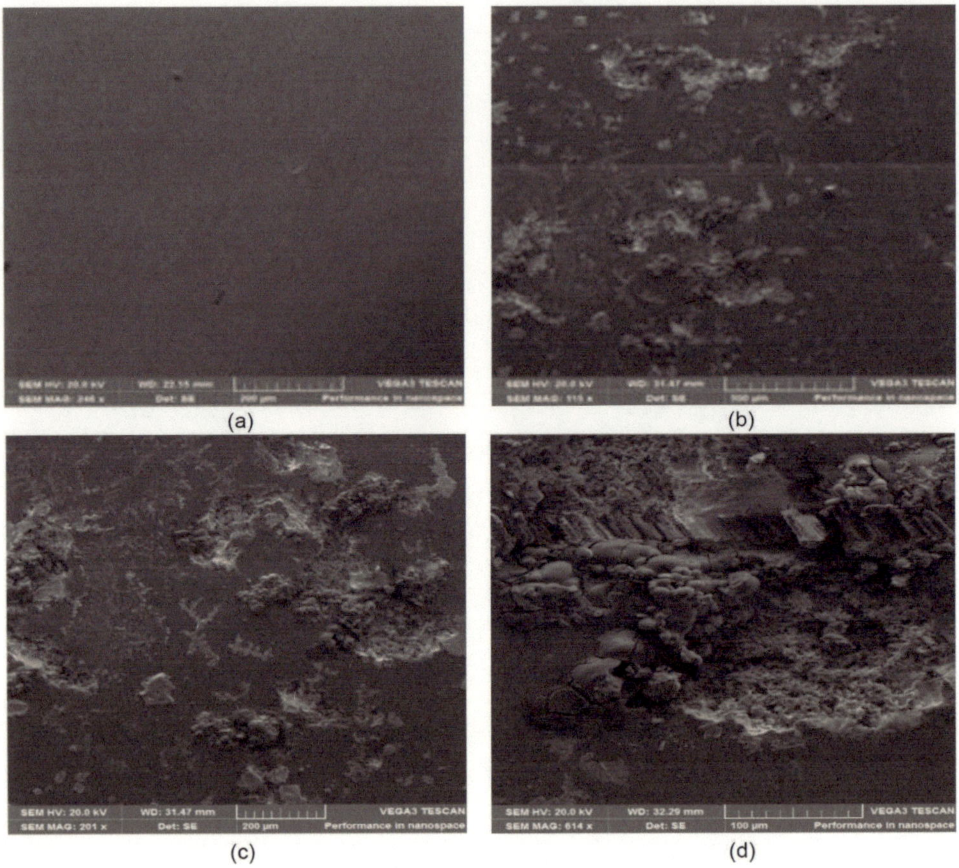

Figure 13.
SEM images of corroded cold-rolled AA8015 sample at $R_a \approx 0.04$ µm. (a) Image before corrosion at 246-x magnification; (b-d) images after corrosion at increased magnifications.

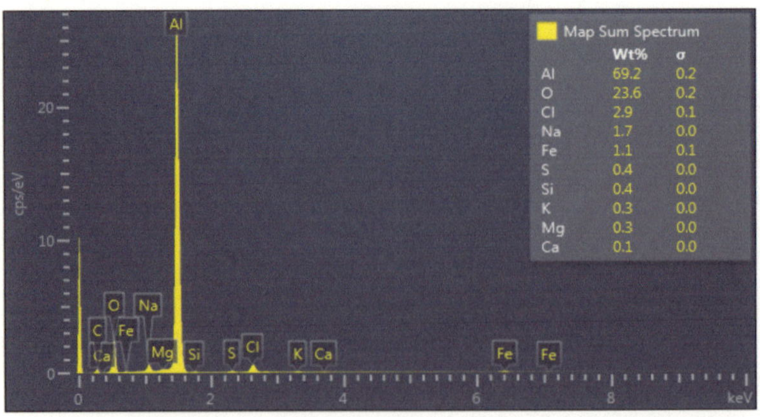

Figure 14.
EDS spectrum showing the elemental composition of the insoluble substrate on the surface.

Additional information

This paper is a revised and expanded version of a paper entitled [Corrosion behaviour of cold-rolled aluminium 8015-alloy in natural sea water at 0.18 μm surface roughness] accepted for presentation at [4th International Conference on Mechanical, Manufacturing and Plant Engineering, Melaka, Malaysia. November 14–15, 2018].

Author details

Olayinka Olaogun[1*], Esther Titilayo Akinlabi[2] and Cynthia Samuel Abima[3]

1 Department of Mechanical Engineering, Kwara State University, Nigeria

2 Department of Mechanical and Construction Engineering, Northumbria University, Newcastle, United Kingdom

3 Department of Mechainical Engineering Science, University of Johannesburg, South Africa

*Address all correspondence to: yinka.olaoluwa@live.com

References

[1] Dieter GE, Kuhn HA, Semiatin SL. Handbook of Workability and Process Design. United States of America: ASM International; 2003

[2] Lewis PR, Reynolds K, Gagg C. Forensic Materials Engineering: Case Studies. New York: CRC Press, Taylor and Francis Group; 2003

[3] Cold Working. Retrieved June 20, 2018. from https://www.corrosionpedia.com/definition/294/cold-working. Copyright 2018 Corrosionpedia Inc.

[4] Hot and Cold Working and the Rolling Process. Retrieved June 20, 2018. from https://www.leonghuat.com/articles/hot&cold.htm.

[5] Cold Working Processes. 2010. Retrieved June 20, 2018. from https://www.totalmateria.com/page.aspx?ID=CheckArticle&site=kts&NM=266.

[6] Schrader FG, Elshennawy KA. Manufacturing Processes & Materials. Michigan, USA: Society of Manufacturing Engineers; 2000

[7] Singh UK, Manish D. Basic Manufacturing Process. 3rd ed. New Delhi: New Age International; 2014

[8] Liu L, Zhang H, Hongyun Bi E, Chang ML. Corrosion behavior of cold-rolled metastable Cr–Mn–Ni–N austenitic stainless steel in acidic NaCl solution. Journal of Materials Research and Technology. 2022;**19**:278-288, ISSN 2238-7854,. DOI: 10.1016/j.jmrt.2022.05.054

[9] Ma G, Wu G, Shi W, Xiang S, Chen Q, Mao X. Effect of cold rolling on the corrosion behavior of Ta-4W alloy in sulphuric acid. Corrosion Science. 2020; **176**:108924, ISSN 0010-938X. DOI: 10.1016/j.corsci.2020.108924

[10] Seshweni MHE, Moloto A, Aribo S, Oke SR, Ige OO, Olubambi PA. Influence of cold and hot rolling on the corrosion behaviour of duplex stainless steels in mine water environment. Materials Today: Proceedings. 2020;**28**(Part 2): 912-915, ISSN 2214-7853. DOI: 10.1016/j.matpr.2019.12.323

[11] Pengfei D, Deng S, Li X. Mikania micrantha extract as a novel inhibitor for the corrosion of cold rolled steel in Cl2HCCOOH solution. Journal of Materials Research and Technology. 2022;**19**:2526-2545, ISSN 2238-7854,. DOI: 10.1016/j.jmrt.2022.06.026

[12] An XL, Zhao H, Chu CL, Dai T, Lu T, Huang ZH, et al. Hall-Petch relationship and corrosion behavior of cold-rolled CoNiFe medium entropy alloy. Journal of Alloys and Compounds. 2019;**807**: 151698, ISSN 0925-8388. DOI: 10.1016/j.jallcom.2019.151698

[13] Asaolu AO, Omotoyinbo JA, Oke SR, Falodun OE, Olubambi PA. Effect of nickel addition on microstructure, tensile and corrosion properties of cold rolled silicon bronze. Materials Today: Proceedings. 2021;**38**(Part 2):1147-1151, ISSN 2214-7853. DOI: 10.1016/j.matpr.2020.07.136

[14] Rao ACU, Vasu V, Govindaraju M, Srinadh KVS. Stress corrosion cracking behaviour of 7xxx aluminium alloys: A literature review. Transactions of Nonferrous Metals Society of China. 2016;**26**(6):1447-1471

[15] Xhanari K, Finšgar M. Organic corrosion inhibitors for aluminium and its alloys in chloride and alkaline solutions: A review. Arabian Journal of Chemistry. 2016;**12**(8):4646-4663

[16] Aribo S, Fakorede A, Ige O, Olubambi P. Erosion-corrosion behaviour of aluminum alloy 6063 hybrid composite. Wear. 2017;**376-377**:608-614

[17] Canepa E, Stifanese R, Merotto L, Traverso P. Corrosion behaviour of aluminium alloys in deep-sea environment: A review and the KM3NeT test results. Marine Structures. 2018;**59**: 271-284

[18] Dai X, Wang H, Ju L, Cheng G, Cong H, Newby BZ. Corrosion of aluminium alloy 2024 caused by aspergillus Niger. International Biodeterioration & Biodegradation. 2016;**115**:1-10

[19] Deepa P, Padmalatha R. Corrosion behaviour of 6063 aluminium alloy in acidic and in alkaline media. Arabian Journal of Chemistry. 2017;**10**:S2244

[20] Donatus U, Thompson GE, Omotoyinbo JA, Alaneme KK, Aribo S, Agbabiaka OG. Corrosion pathways in aluminium alloys. Transactions of Nonferrous Metals Society of China. 2017;**27**(1):55-62

[21] Zhang X, Zhou X, Hashimoto T, Liu B. Localized corrosion in AA2024-T351 aluminium alloy: Transition from intergranular corrosion to crystallographic pitting. Materials Characterization. 2017;**130**:230-236

[22] Kim S, Jang S, Han M, Park J, Jeong J, Chong S. Mechanical and electrochemical characteristics in sea water of 5052-O aluminium alloy for ship. Transactions of Nonferrous Metals Society of China. 2013;**23**(3):636-641

[23] Krogstad HN, Johnsen R. Corrosion properties of nickel-aluminium bronze in natural seawater—Effect of galvanic coupling to UNS S31603. Corrosion Science. 2017;**121**:43-56

[24] Liu B, Zhang X, Zhou X, Hashimoto T, Wang J. The corrosion behaviour of machined AA7150-T651 aluminium alloy. Corrosion Science. 2017;**126**:265-271

[25] Zhang SL, Zhang FD, Liu YX, Liu H, Liu Z, Suebka C, et al. Corrosion behaviour of laser-cleaned AA7024 aluminium alloy. Applied Surface Science. 2018;**435**:452-461

[26] ASTM G1. Standard Practice for Preparing, Cleaning, and Evaluating Corrosion Test Specimens. Philadelphia: ASTM International; 2011. DOI: 10.1520/G0001-03R11

[27] Struers' Application Notes: Metallographic Preparation of Aluminium and Aluminium Alloys, Sponsored by Struers Inc. [online] https://www.struers.com/-/media/Struers-media-library/Materials/Application-reports/AN-11_Aluminium_2015_ENG.pdf [Accessed: 18 March 2016]

[28] Water Condition and Purification. 2005. Accessed from https://www.lenntech.com/composition-seawater.htm on 8 May 2018

[29] ASTM G5-14: Standard Reference Test Method for Making Potentiodynamic Anodic Polarization Measurements American Society for Testing and Materials (Filadelfia, Pennsylvania). ASTM; 2014. Available from: https://books.google.com.ng/books?id=tWpcswEACAAJ

[30] ASTM G69, Standard Test Method for Measurement of Corrosion Potentials of Aluminum Alloys, West Conshohocken, PA, 2012. DOI: 10.1520/G0069-12

Chapter 11

Experimental Investigations on Advancements of Aluminum Alloys with Friction Stir Process

Bazani Shaik, M. Muralidhara Rao, G. Harinath Gowd, B. Durga Prasad and J. Ranga

Abstract

Friction stir processing is a very promising method widely joining varieties of metals in other relatively marine, shipbuilding, automotive industries, aeronautical, and heavy machinery industries due to the following advantages, such as, low porosity, less tendency to cracking, and fewer defects. Research investigates the mechanical properties for input parameters such as welding speed, rotational speed, tilt angle, and axial force, and output parameters such as tensile strength, microhardness on advancements of aluminum alloys by using friction stir processing based on cost. Taguchi L9 used for the carrying research on experiments with trailing on parent materials in different ranges of input responses on welding speed is 60 mm/min, rotational speed 1250 rpm, tilt angle 3°, and axial force of 12 KN output responses tensile strength are 167 MPa measured on the basis of ASTM on specimens and analysis for carrying and using design of experiments and mathematical modeling, the relations with empirical process useful for the development for automated design.

Keywords: aeronautical industries, mechanical properties, advancements of aluminum alloys, friction stir process, welding

1. Introduction

The friction stir welding process is currently very useful for ship manufacturing and industry-oriented aircraft and automotive for butt, lap with spot-on dissimilar joining of applicability Al-alloys and other materials of Mg-alloys, the production of mass of light transportation systems and fuel consumption has significantly reduced [1]. Studied resistance of ironing with process aluminum alloys are increased to improve the silicon oxide nanoparticles for the limit of iron [2]. Studied mechanical properties and microstructural evaluation of AZ31B of sheets has 3 mm thickness welded of optimum conditions. The material of workpieces for joining friction stir processing with tool is shown in **Figure 1** [3, 4]. Studies on tempered steel with quench property are feasible of tensile strength 1635 Mpa and research focus of different types of high carbon steels and medium are accepted successfully of friction stir welds. Joining of Al6061 or NiTip composite with the distribution of homogeneous

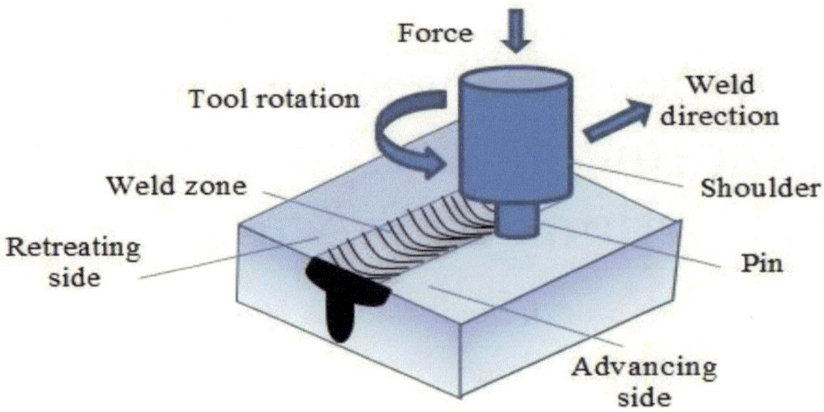

Figure 1.
Friction stir welding process.

particles without product interface reaction is prepared successfully by friction stir processing took place combination of good damping with thermal physical properties on the treatment of heat process in the composite [5, 6]. Al–Li 2099 T86 of stress corrosion cracking applications and [7, 8] developments of new alloy in aircraft industries are identified aluminum–lithium alloys with the [9, 10] substitute of high strength aluminum alloys on spacecraft manufacturing and launchers. The properties of strength, toughness, and [11, 12] stiffness are adopted with aluminum alloys. The aluminum–lithium alloys advanced took place with stress corrosion [13, 14] cracking on structural space applications. The parameters used for welding have [15, 16] cohesive bands and circular shapes and path studied of tool intention.

2. Materials and methods

The friction stir process mainly involves the basic need for materials and methods influences by welding of dissimilar AA7075T651 and AA6082T651 with having thickness of 6 mm and by using advanced numerically controlled stir process are carried out experiments on the basis of lot of literature survey and trail error methods on input parameters varying with proportionate condition done at Annamalai university. Chemical compositions with base material are shown in **Table 1**. The specimens of the plate taken dimensions on the basis of gap is 100 mm × 50 mm × 6 mm. The dimensions cut by the edges with smooth areas to do easily joining process of butt welding for the two dissimilar aluminum alloys are placed advancing side and retreating side are shown in **Figure 2** for the fixed clamps will be adjusted for specimens.

Elements	Si	Fe	Cu	Mn	Mg	Cr	Ni	Zn	Ti	Al
Al7075-T651	0.12	0.2	1.4	0.63	2.53	0.2	0.004	5.62	0.03	89.26
Al6082-T651	1.05	0.26	0.04	0.68	0.8	0.1	0.005	0.02	0.01	97.03

Table 1.
The chemical compositions of AA7075T651 and AA6082T651.

Figure 2.
Weld position of dissimilar aluminum alloys of friction stir welding.

The designed tool with advanced condition material taken as M2-Grade SHSS tool diameter of the shoulder is 18 mm and length of the probe is 6 mm. After the friction stir processing, the weld zone appears perfectly, for the testing of the welding specimens are taken as standards of ASTM E8 and tensile test specimens before shown in **Figure 3** and specimens after testing are shown in **Figure 4**. The combination and particular diameter of standard specimens are taken for the impact strength shown in **Figure 5**. The AA7075T651 advancing side and AA6082T651 in retreating side to have the proper joining of materials and for the improvement of mechanical properties.

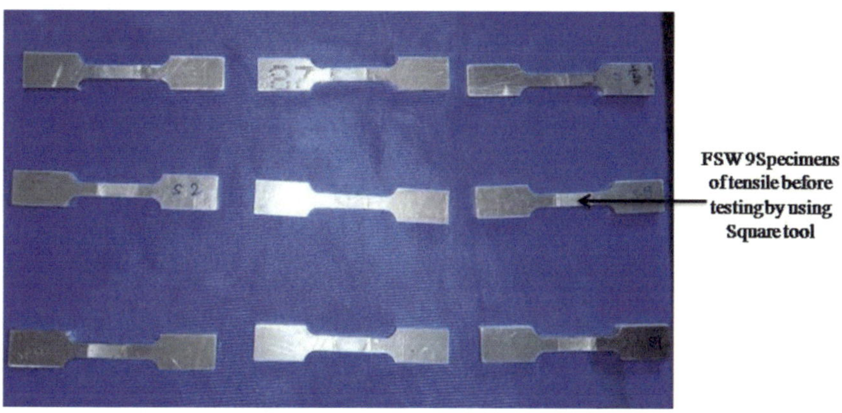

Figure 3.
Specimens of tensile test before testing with ASTM E8.

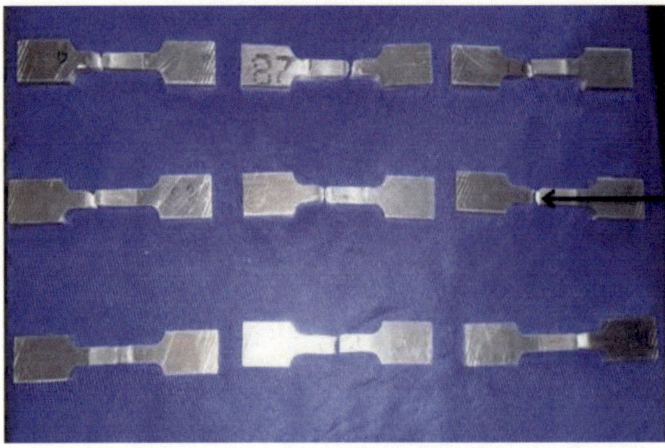

Figure 4.
Specimens of tensile test after testing.

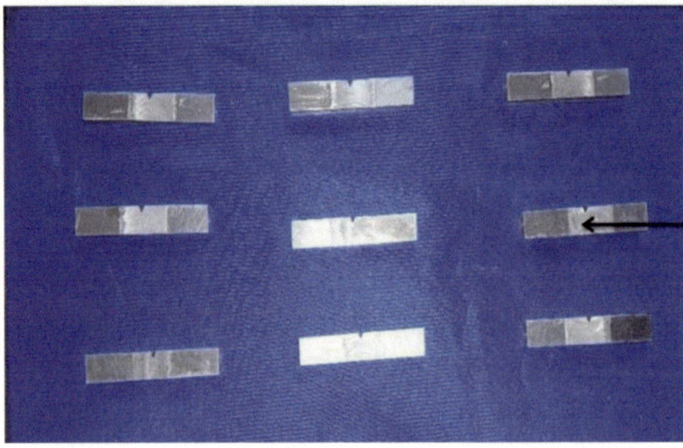

Figure 5.
Specimens of impact test.

S. no	Parameters	Notation	Unit	Levels		
				1	2	3
1	Welding speed	WS	mm/min	40	50	60
2	Rotational speed	RS	rpm	1150	1250	1350
3	Axial force	AF	KN	9	10.5	11

Table 2.
Input variables for actual and coded.

The advanced methodology applied for different parameters to obtain easy way of influencing the properties of mechanical by using dissimilar welding of notations and units are described in **Table 2** and experimental design of Taguchi model input parameters and output parameters is shown in **Table 3**.

Exp. no	Input process parameters				Output responses		
	Rotational speed (rpm)	Welding speed (mm/min)	Tilt angle (degree)	Axial force (KN)	Tensile strength (MPa)	Impact strength (J)	Elongation (%)
1	1150	40	1	10	162.00	10.55	9.60
2	1150	50	2	11	158.99	10.31	9.41
3	1150	60	3	12	155.00	9.00	8.50
4	1250	40	2	12	171.00	12.20	10.80
5	1250	50	3	10	164.99	11.16	10.05
6	1250	60	1	11	158.00	9.30	6.78
7	1350	40	3	11	174.99	13.10	12.15
8	1350	50	1	12	173.00	13.03	11.25
9	1350	60	2	10	167.00	11.30	10.10

Table 3.
Experimental design of Taguchi model.

3. Design of expert

The design of experts in series with the test for the researcher useful for changes in input variables on a processor system is shown in **Figure 6** due to the effect of variables of responses measured. The applicability of computer simulation models and physical on the factorial designs took place sensitively for the estimation of the combination of effect for two or more factors.

The design of experiments and methods of the traditional difference taken place approach in a better way of values on variables of parallel and it does not cover the main effects on the variables on the different interactions and the possibility of approach for identifying optimal values on the variables of combination with

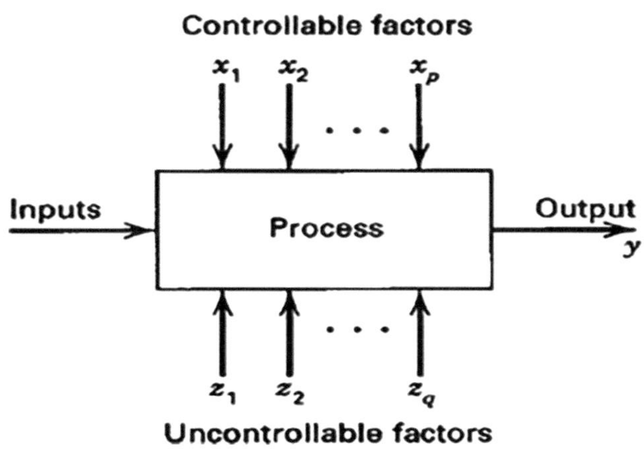

Figure 6.
Process model of the design of expert.

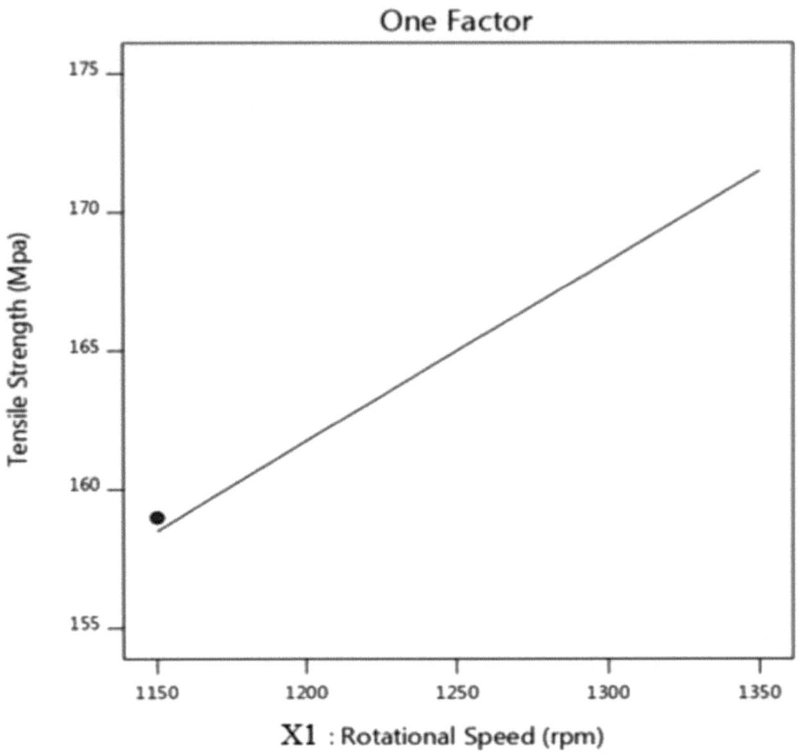

Figure 7.
Influence of rotational speed on tensile strength.

experimental runs. The design of experiments is carried out in four phases: Screening, Planning, Optimization, and Verification.

The influence of rotational speed on tensile strength has increased based on the tool welding speed varies the strength with respect to the elongation has improved to the maximum extent depends on the rotational speed. **Figure 7** shows the increases in rotational speed depends on the heat increases at the welding zone area. The friction coefficient decreases with the melting condition. The friction stir process region intricate the fine particles will be distributed in the uniform portion. The effect of tool stirred the position on the flow of metal optimum depends on the increase of tensile strength.

The percentage of elongation along transverse direction obtained from the tensile test plotted against the welding speed. The plates (**Figure 8**) shows welded with a rotational speed of 1250 rpm and weld speed of 40 mm/min. While the plates were welded at 1150 rpm and 60 mm/min. The influence shows the properties of higher heat input on the basis of influenced elongation.

The influence of tilt angle on tensile strength (**Figure 9**) shows the manner of the position at the bottom area of the welded part and it will increase the position of tool speed with respect to the material and designed shoulder based. The region of the position will make difference between the tool changes the yield strength to improve the microstructure with ductility.

The influence of axial force on tensile strength (**Figure 10**) shows the significance of friction stir processing at the joining area. The joint took place in the position

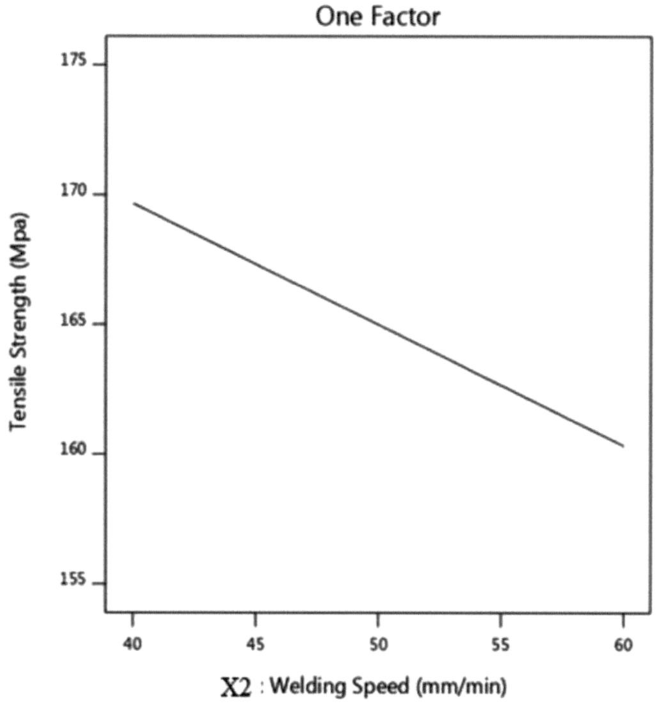

Figure 8.
Influence of welding speed on tensile strength.

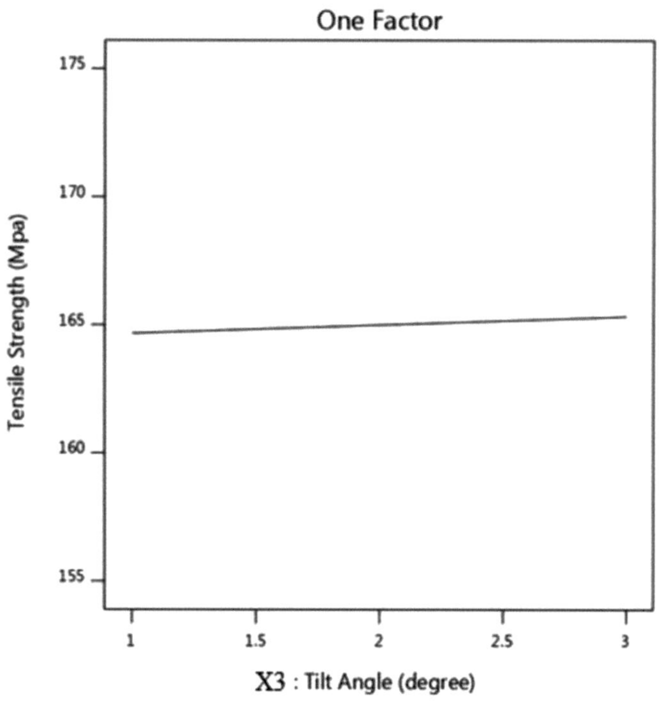

Figure 9.
Influence of tilt angle on tensile strength.

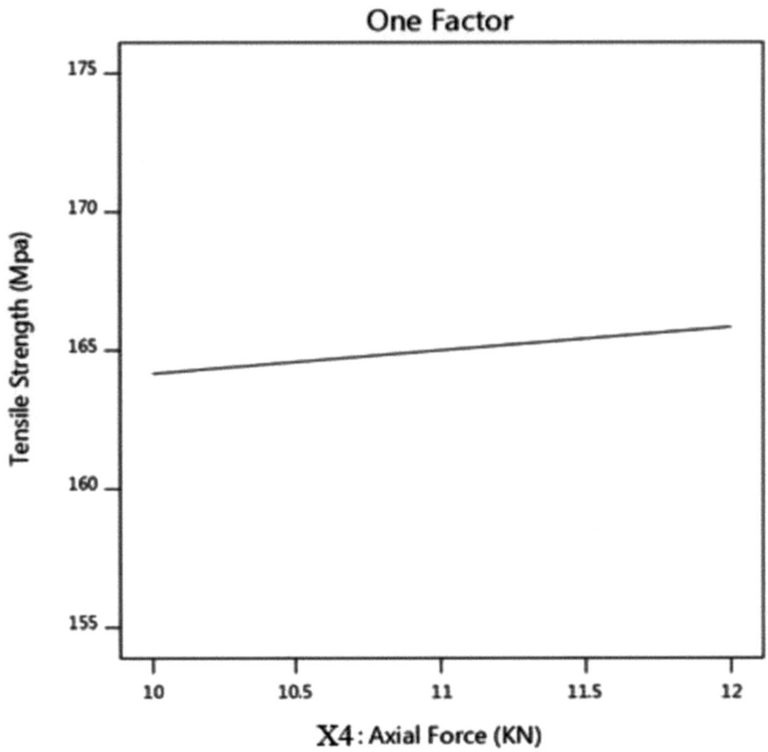

Figure 10.
Influence of axial force on tensile strength.

of rotational speed is 1250 rpm and tensile strength 164.99 MPa and the welding speed takes the major role due to increasing of force is 12 KN has the strength will be superior at the position of part counter.

The influence of rotational speed on impact strength produces **Figure 11** shows the frictional heat required to plasticize the material and also the effect of proper mixing of the dissimilar alloys. The changes in the position of the part speed will be low and have good mechanical properties at the welding speed is higher.

The influence of welding speed on impact strength shows **Figure 12** maintains the region with the center point of the notch makes the higher energy in order to analyze the impact energy at an instant with the increasing of welding speed 60 mm/min and impact energy of the notch shows 9.2 J.

The influence of tilt angle on impact strength shows **Figure 13** increases of the impact energy with 11 J with respect to the tilt angle 3° will be maximum of increasing tool tilt angle.

The influence with axial force on impact strength **Figure 14** shows the tool stirring action plays a major role of the part to increase the rotational speed with the resultant of the weld area. The surfaces that occur groove condition because insufficient material will be visible. The zone of the weld part decreases with rotational speed due to the effect of distribution with temperature at the area of weld zone.

The influence of rotational speed on elongation shows in **Figure 15** with the increase of rotational speed on the higher input of heat. The position of the tool will be the friction decreases with the heat input condition. The friction stir processing is

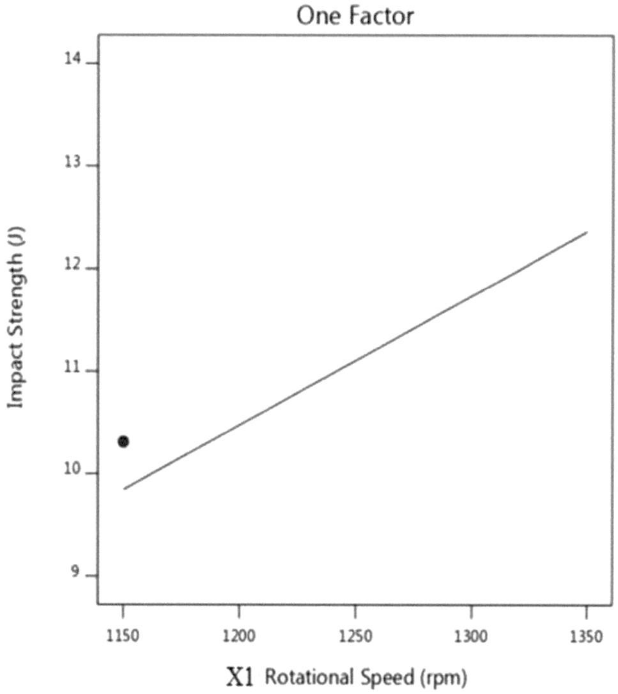

Figure 11.
Influence of rotational speed on impact strength.

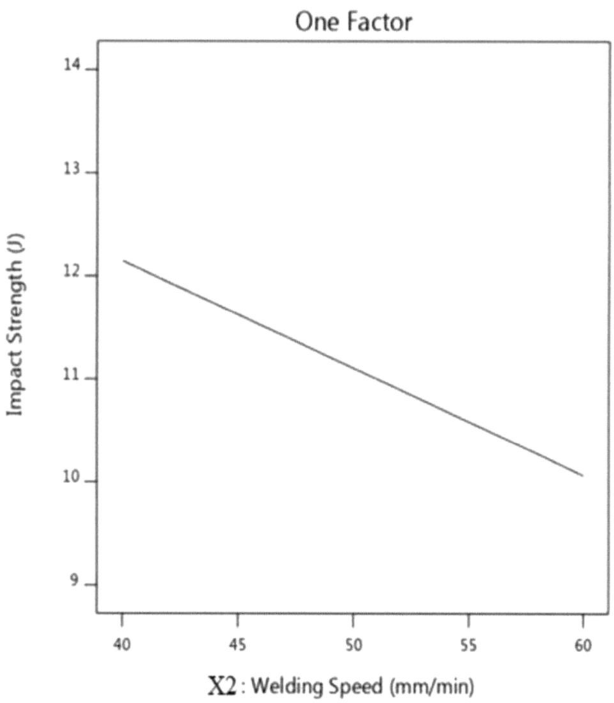

Figure 12.
Influence of welding speed on impact strength.

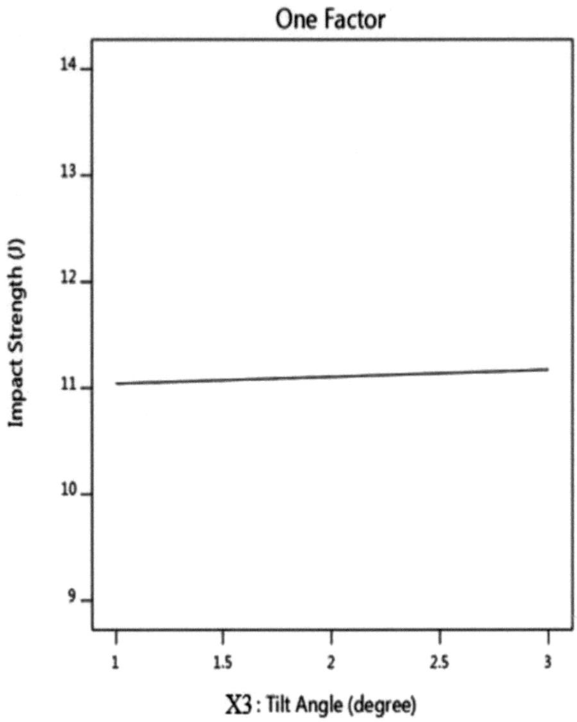

Figure 13.
Influence of tilt angle on impact strength.

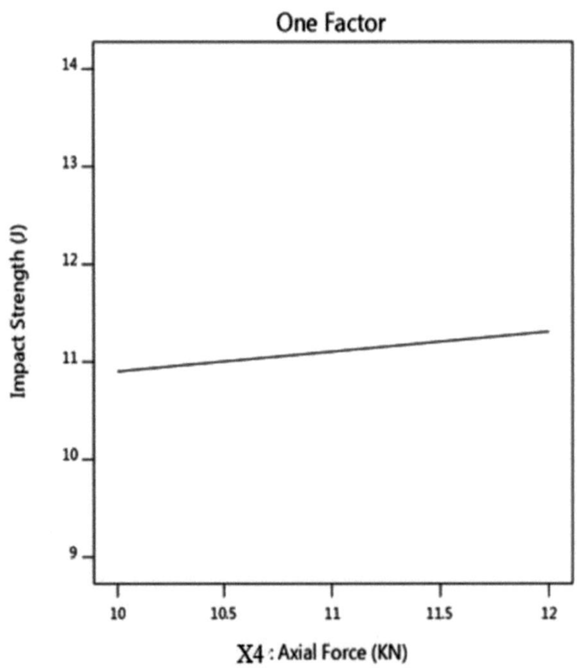

Figure 14.
Influence of axial force on impact strength.

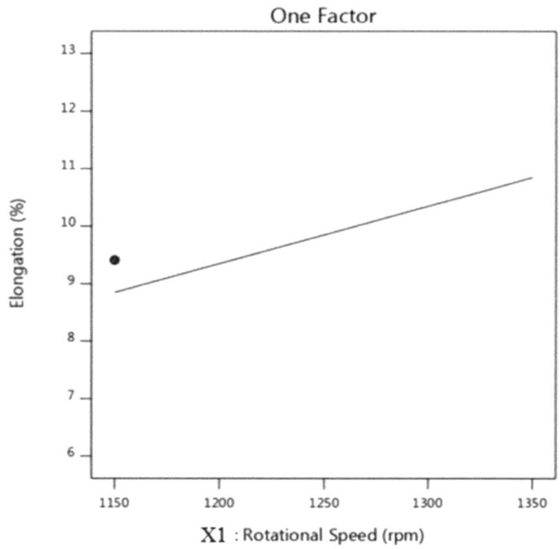

Figure 15.
Influence of rotational speed on elongation.

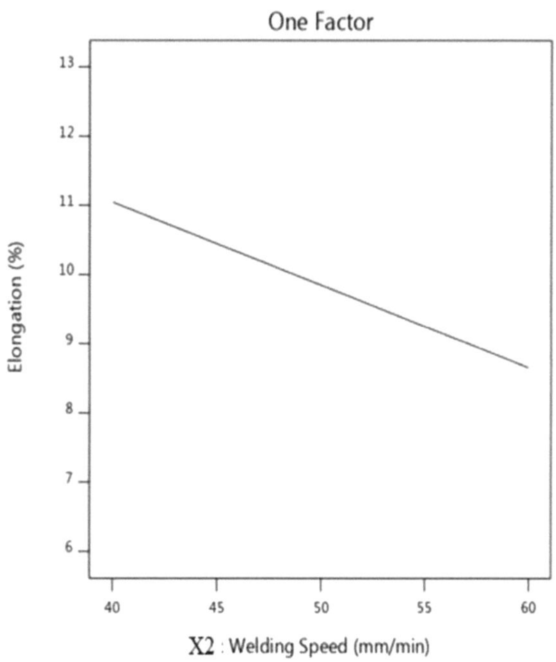

Figure 16.
Influence with welding speed on elongation.

the best condition for the optimized region on the fine particles with the distribution of uniform.

The percentage of elongation along transverse direction obtained from the tensile test plotted against the welding speed. **Figure 16** shows plates welded with rotational speed is 1250 rpm and weld speed of 40 mm/min. While plates welded rotational speed is 1150 rpm and welding speed 60 mm/min. The proportion area influences the heat input due to the elongation of 11.25%.

The influence of tilt angle on elongation shown in **Figure 17** is the position of tool depends on the material adjustment at the shoulder region of the part condition varies with the improvement condition in a friendly environment at the joining portion of

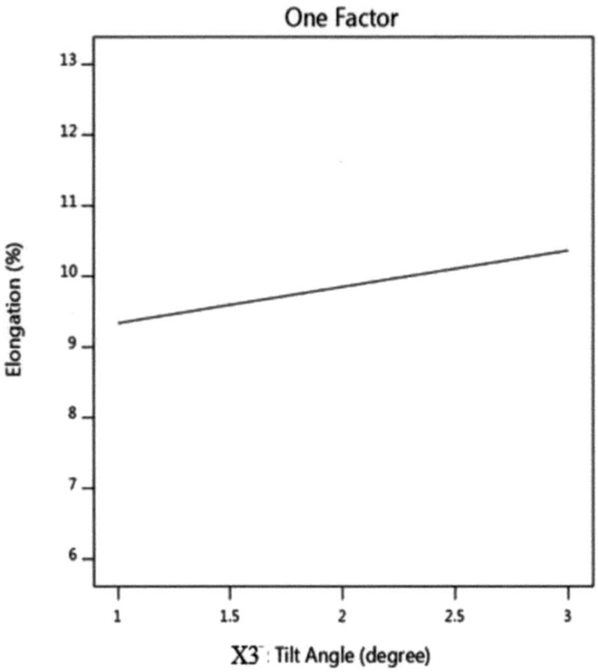

Figure 17.
Influence of tilt angle on elongation.

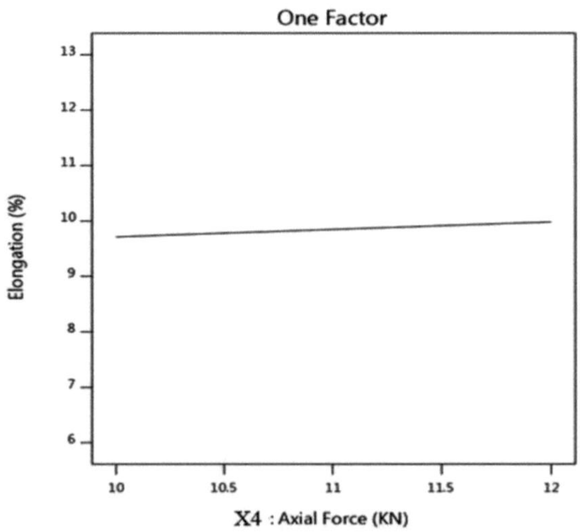

Figure 18.
Influence of axial force on elongation.

the yield works due to the microstructure will give perfect condition in the region of the part due to tilt angle maximum 3° and elongation of 9.7%.

The influence of axial force on elongation shows in **Figure 18** with the flow of zone part due to higher heat input it occurs at the probe area. The tool pin changes the position in order to move the actual flow of material to control the plastic deformation easily. The shoulder will be the major portion force will increase the depth level of plunge working the linear position. The axial force increases the due to increase of the pressure at a higher extent the shoulder area will be stirred normal position easily.

4. Conclusions

The present investigation shows the aluminum alloys with the application of Taguchi design of experiments helped us in conducting the experiments in an effective manner without losing accuracy. Two-dimensional plots are plotted between the input process parameter and the output responses using Design-Expert software. The tensile strength is increasing with the increase in rotational speed and the axial force values and the tensile strength is decreasing with the increase in the weld speeds. The impact strength increases, when there is an increase in the values of rotational speed and axial force. Whereas the impact strength tends to decrease with the increase in the weld speeds. The elongation also increases with the increase in rotational speed and axial force. The results presented in the work are analyzed on the basis of analysis process conducted with microstructures with different zones on thermo mechanical treatment zone has higher plasticity due to eutectic constituents Cu–Al precipitation on rolled condition and parent metal has rolled temper condition.

Author details

Bazani Shaik[1*], M. Muralidhara Rao[1], G. Harinath Gowd[2], B. Durga Prasad[3] and J. Ranga[1]

1 Ramachandra College of Engineering, Eluru, Andhra Pradesh, India

2 Madanapalle Institute of Technology and Sciences, Madanapalle, Andhra Pradesh, India

3 Jawaharlal Nehru Technological University, AnantapuramuAndhra Pradesh, India

*Address all correspondence to: drbazanishaik@rcee.ac.in

References

[1] Dorbane A, Ayoub G, Mansoor B, Hamade RF, Kridli G, Shabadi R, Imad A. "Microstructural observations and tensile fracture behavior of FSW twin-roll cast AZ31 Mg sheets". Materials Science and Engineering A. 2016;**649**:190-200-September 2015

[2] El-Batahgy A, Miura T, Ueji R, Hidetoshi F. "Investigation into the feasibility of FSW process for welding 1600 MPa quenched and tempered steel" Materials Science and Engineering A. 2016;**651**:904-913-January 2016

[3] Ni DR, Wang JJ, Ma ZY. Shape memory effect, thermal expansion and damping property of friction stir processed NiTip/Al composite, Journal of Materials Science and Technology. 2015. http://dx.doi.org/doi: 10.1016/j.jmst.2015.12.013-October 2015

[4] Goebel J, Ghidini T, Graham AJ. "Stress-corrosion cracking characterisation of the advanced aerospace Al-Li 2099-T86 alloy", Materials Science and Engineering A. http://dx.doi.org/10.1016/j.msea.2016.07.013 July 2016

[5] Ma ZY, Feng AH, Chen DL, Shen J. Recent advances in friction stir welding/processing of aluminum alloys: Microstructural evolution and mechanical properties. Critical Reviews in Solid State and Materials Sciences. 2017. DOI: 10.1080/10408436.2017.1358145

[6] Dragatogiannis DA, Koumoulos EP, Kartsonakis I, Pantelis DI, Karakizis PN, Charitidis CA. Dissimilar friction stir welding between 5083 and 6082 Al alloys reinforced with Tic nanoparticles. Materials and Manufacturing Processes. 2015. DOI: 10.1080/10426914.2015.1103856

[7] Patel AR, Drupal J, et al. Investigation of mechanical properties for hybrid joint of aluminium to polymer using friction stir welding. Materials Today: Proceedings. 2018;**5**:4242-4249

[8] Das U, Toppo V. Effect of tool rotational speed on temperature and impact strength of friction stir welded joint of two dissimilar aluminum alloys. Materials Today: Proceedings. 2018;**5**:6170-6175

[9] Dalwadi CG, Patel AR, Kapopara JM, Kotadiya DJ, Patel ND, Rena HG. Examination of mechanical properties for dissimilar friction stir welded joint of Al alloy (AA-6061) to PMMA (acrylic). Materials Today: Proceedings. 2018;**5**:4761-4765

[10] Bazani Shaik, G. Harinath Gowd and B. Durga Prasad, Investigations and optimization of friction stir welding process to improve microstructures of aluminium alloys, Taylor & Francis Online, May 2019. https://doi.org/10.1080/23311916.2019.1616373.

[11] Shaik B, Gowd GH, Prasad BD. An optimization and investigations of mechanical properties and microstructures on friction stir welding of aluminium alloys. International Journal of Mechanical and Production Engineering Research and Development. Feb 2019;**9**(1):227-240. ISSN(P): 2249-6890; ISSN(E): 2249-8001

[12] Shaik B, Gowd GH, Prasad BD. Durga Prasad, Investigations on friction stir welding process to optimize the multi responses using GRA method, International Journal of Mechanical Engineering and Technology (IJMET). March 2019;**10**(3):341-352. ISSN Print: 0976-6340 and ISSN Online: 0976-6359

[13] Shaik B, Gowd GH, Prasad BD. Experimental investigations on friction stir welding process to join aluminum alloys. International Journal of Applied Engineering Research ISSN 0973-4562. 2018;**13**(15):12331-12339

[14] Shaik B, Gowd GH, Prasad BD. Durga Prasad, Parametric investigations on friction stir welding of aluminium alloys. In Emerging Trends in Mechanical Engineering, Lecture Notes in Mechanical Engineering. ISSN 2195-4364, ISBN 978-981-32-9931-3. https://doi.org/10.1007/978-981-32-9931-3_33

[15] Shaik B, Gowd GH, Prasad BD. Experimental and parametric studies with friction stir welding on aluminium alloys, Materials Today: Proceedings. 2019;**19**:372-379. https://doi.org/10.1016/j.matpr.2019.07.615

[16] Shaik B, Gowd H, Durga Prasad B, Siddik P. Investigations on microstructures by using friction stir processing. Intelligent Manufacturing and Energy Sustainability, Smart Innovation, Systems and Technologies. In: Reddy ANR, Deepak M, Margarita N, Favorskaya, Milan S, Suresh CS, editors. Singapore; 1st ed. 2020 edition. Springer Verlag; 2022. pp. 539-548